Limited
Energy Systems

A Guide to Alarm, Signaling,
Remote-Control, and
Communications Circuits

Noel Williams

NFPA
Quincy, Massachusetts

Product Manager: Brad Gray
Editorial-Production Services: Publishers' Design and Production Services, Inc.
Composition: Publishers' Design and Production Services, Inc.
Cover Design: Cameron Incorporated
Manufacturing Manager: Ellen Glisker
Printer: Edwards Brothers

NFPA No.: LES-02
ISBN: 0-87765-519-7

Library of Congress Control No.: 2001099202

Printed in the United States of America
02 03 04 05 06   5 4 3 2 1

Contents

Preface

WHY IS THIS BOOK NEEDED?

Remote control, signaling, and communications circuits have seen a dramatic increase in applications and use in recent years. Much of this increase has been due to the expanding role of computer and Internet-related systems. Fax machines, security systems, fire alarm systems, home automation, and other uses have also increased the need for specialized wiring. This expanding market has drawn many new users, designers, installers, contractors, and manufacturers into limited energy installations. In some cases, what was once done by large communications utilities is now being done by electrical or specialty contractors. Some of the new players are more familiar with accepted installation practices and safety requirements than others.

As a result of the increasing numbers of applications and installations of limited energy circuits, many building and fire safety officials have taken an increased interest in these installations. One reason for their interest is the proliferation of cables related to limited energy systems. Not only are the numbers of such cables increasing due to increased use, but in many cases, new technology requires that new types of cables be installed, and often the existing cables have been abandoned in place. Abandoned cables represent an increased fire load with no function. Some fire officials and insurance inspectors are very much concerned with these conditions. These cabling systems have been largely ignored in the past, partly because the cabling was often installed after buildings and final inspections were completed. Now some experienced installers are finding their installations subject to a new scrutiny to which they are not accustomed. The *NEC* has responded to changes in the industry as well as to specific concerns and now requires that abandoned cables be removed. The *NEC* also applies new support and installation requirements for the limited energy cables.

Not all of the increased interest is based on safety considerations. For example, many electrical contractors have started installing communications and data wiring, and in some cases are in direct competition with utilities who formerly did much of this wiring. This book is not intended to address such political, labor, or market-related issues. Nor is this book concerned with the satisfactory performance or operation of any particular system or any product-specific recommendations. Rather, this book is intended as a resource for those who wish to ensure that limited energy equipment and installations are safe and code-compliant.

WHO SHOULD USE THIS BOOK?

This book is directed at all users of the *NEC* who design, install, or inspect limited energy systems. Limited energy systems may include those used for remote control, signaling, data, fire alarm, communications, video, or other uses. The function of such systems is generally not the major consideration of the *Code* except in cases in which a failure of the limited energy circuit itself creates a safety hazard. While reliability of such systems may be critical to business operations or other issues of "convenience," in most cases, the functional failure of a limited energy system such as a doorbell or thermostat does not produce a fire or life-safety hazard. In general, the *NEC* is directed at safety issues, or as stated in Section 90-1, the *NEC* is concerned with the protection of persons and property from the hazards associated with the use of electricity. The same section states that compliance with the *Code* will not necessarily ensure a convenient or adequate installation. Also, like the *NEC*, this book is not intended as a design manual or specification. Adequacy, convenience, and the performance of desired functions are left to designers.

Introduction

OVERVIEW OF LIMITED ENERGY APPLICATIONS

Limited energy systems are found in all types of occupancies and cover a wide range of applications, including some remote control and signaling circuits, fire alarm wiring utilizing circuits of less than 120 volts, and most communications circuits. However, within each category, some applications may not be limited energy. For example, remote control of motors often uses 120-volt circuits, which exceed the power levels allowed for limited energy circuits. Similarly, some complete fire alarm systems and certain parts of other fire alarm systems are not power limited. Nevertheless, most occupancies do contain some limited energy circuits and systems. These systems are common even in residential occupancies in circuits for doorbells, thermostats, garage door openers, telephones, cable or satellite TV, and, perhaps, security, fire, intercom, and some audio systems. In newer construction of all types, systems for lighting control, energy management, temperature control, computer networks, communications, and automation are common, and these systems usually are based on or include limited energy circuits.

The circuits and systems covered by this book are primarily those derived from power sources that have inherent power limitations. In fact, the terms *energy limited*, *limited energy*, and *power limited* are considered interchangeable. Although these systems are often loosely referred to as *low voltage,* voltage limitations alone do not allow the use of the special wiring methods permitted for energy-limited circuits (see Chapter 2).

This book primarily covers those systems and circuits using power sources that limit both voltage and current to minimize fire or shock hazards. Because of the reduced hazards, special wiring methods and materials are permitted. The proper identification and installation of these systems and the proper use of the special wiring methods are the main goals of this book. However, this book does cover Class 1 circuits and some other nonpower-limited circuits where they are closely related to power-limited circuits. This book also covers optical fiber cables. Although optical fibers are not themselves electrical, they are often used with or for similar purposes as limited energy circuits.

USING THIS BOOK

This book is organized by type of system rather than by type of requirement. Many users of the *National Electrical Code*® (*NEC*®) specialize in certain types of installations. Although the rules for various types of limited energy systems are similar, there are also subtle differences. Rather than trying to distinguish those differences in requirements for different systems, this book treats each type of system separately. This approach to the subject is intended to allow the designer, installer, or inspector to concentrate only on individual areas of interest.

Chapter 8 of the *NEC* is titled "Communications Systems," but four different types of circuits are included under this general heading. Communications systems are treated the same way in Chapter 6 of this book; that is, as one system that includes four different types of circuits.

Many users of this book have a specific area of interest. Readers who are primarily interested in one type of circuit and have limited experience with the *NEC* should read Chapters 1 and 2 and then Chapter 3, 4, 5, or 6, depending on their area of interest. Readers with a particular interest area and a thorough understanding of *Code* conventions and organization may want to concentrate only on the chapter that addresses their area of interest. Other users, such as inspectors, as well as many designers and installers, will encounter occasional applications in all areas. They may need to review each type of system individually.

As noted, the various types of limited energy circuits have many similarities. However, each type of system is covered individually in this book to enable comparing and contrasting of requirements for the various types by organizing the coverage of each system in the same overall manner. In general, this book is organized in the same way as the *NEC*; that is, the requirements are ordered from the most general to the most specific. In Chapter 1 we examine the general organization and conventions of the *NEC*. In Chapter 2 we take a look at limited energy applications and distinguish these applications from other low voltage applications. Finally, in Chapters 3 through 7 we discuss common types of limited energy applications in the order in which they are covered in the *NEC*, with the exception of the closed-loop and programed power distribution systems covered by Article 780. This type of system represents a specific technology or method of combining power distribution with control wiring. We do not cover this type of system in detail. Users and inspectors of these systems should rely primarily on the instructions included in the listing and labeling of the systems.

The ordering of the coverage of systems by type matches the ordering in the *NEC*. Similarly, within the discussions of each system, the coverage generally follows the arrangement of the corresponding *NEC* articles. This ordering is intended to make this book easy to use as a study aid in conjunction with the *Code* text included in each chapter.

Since this book does not repeat all of the *NEC* requirements or cover every possible variation or installation, it is must be used in conjunction with the *NEC* if it is to be applied to actual installations. This book is intended to help users of the *Code* to understand the reasons for and the intent behind the *Code* requirements.

Application of the *National Electrical Code*®

An understanding of the purpose and organization of the *National Electrical Code*® (*NEC*®) is critical to any user who wants to properly interpret and apply its requirements. This understanding is even more critical to those who work in the specialty areas of the *NEC* in which limited energy circuits are covered by special rules that may dramatically alter the usual application of the *Code*. This chapter contains brief explanations of the purpose and organization of the *Code* and of how the ordinary rules that apply to power and lighting circuits are modified or are otherwise different for limited energy circuits.

PURPOSE OF THE *NEC*

Section 90.1 of the *NEC* states that the purpose of the *Code* is "the practical safeguarding of persons and property from hazards arising from the use of electricity." Many of the limited energy circuits covered in this book do not present significant risks of fire or shock. In some cases a risk of shock is present, but the circuit involved would not normally be capable of starting a fire. In other cases the risks addressed have little to do with inherent risks of a system, but are concerned with the combustible materials that may be used in the installation of a circuit. Generally, the *NEC* is not concerned with satisfactory performance of a system, unless operation of the system itself is a safety requirement, such as the operation of an emergency power system or a fire pump. Therefore, some rules that an installer may have to follow to ensure that a circuit operates properly are not rules that are contained in the *NEC*. The *NEC* usually accommodates those installation issues that are necessary for a system to function properly while establishing only those installation requirements that are necessary to safety.

CODE ORGANIZATION AND HIERARCHY OF RULES

The *NEC* and many of the articles in the *NEC* share the same overall organization from the most general rules to the most specific (see Figure 1.1). In fact, the first part of many of the articles is titled

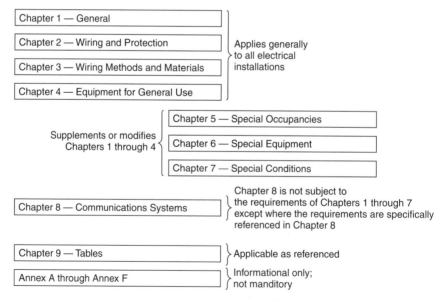

FIGURE 1.1 Organization of the 2002 edition of the *National Electrical Code®*. (Source: *National Electrical Code®*, NFPA, 2002, Figure 90.3)

"General" and covers those rules that are most encompassing. As we move through the *Code* or through an article, the rules become more specific. Thus, Article 100 covers definitions of terms that apply across the *Code*, and Article 110 provides very general rules that apply to all types of installations. The articles in Chapter 7 of the *Code*, in contrast, are very specific in covering only certain types of circuits and installations.

This organization of most general to most specific is made explicit in Section 90.3, which states that the first four chapters of the *Code* apply generally and that Chapters 5, 6, and 7 modify or supplement these general rules. In some cases the modifications provide significant restrictions that narrow the range of options for a particular installation. Certainly this is the case with those articles in Chapter 5 of the *Code* that deal with hazardous (classified) locations. The wiring methods that can be used in such occupancies are only a subset of all the methods available generally. In comparison, the rules for some special equipment, such as the welding equipment covered by Article 630, may need to be liberalized somewhat to deal with the characteristics of the equipment. Thus, a user of welders and the circuits supplying welders gets significant leeway in the selection of overcurrent protection for such circuits and equipment. In this case, the rules regarding welder circuits modify the rules of Article 240 in Chapter 2.

In effect, then, Section 90.3 provides the permission for the rules in Chapters 5, 6, and 7 to change the rules in the first four chapters. This permission and an understanding of the *Code* hierarchy is very important to anyone applying the *NEC* to limited energy circuits.

For communications circuits and the other systems covered in Chapter 8 of the *Code*, nothing in the first four chapters applies except where those rules in the first four chapters are specifically referenced in Chapter 8. This fact is stated in Section 90.3 and illustrated in Figure 1.1. Article 800 does refer to Chapter 3 of the *Code* for the proper installation of a raceway used for communications wiring, and it refers to Section 110.3(B) that requires compliance with listing and labeling instructions. However, as an example, the general requirements for working space for and access to equipment are not referenced in Article 800, so those rules do not apply to communications installations. These issues are covered more fully in Chapter 6 of this book.

GROUNDING REQUIREMENTS

Grounding requirements in the *NEC* can be divided into two general types: systems or circuits that are required to be grounded and equipment or enclosures that are required to be grounded. The general rules of Article 250 require that certain circuits and systems and most enclosures and equipment be grounded. These requirements may be modified for the limited energy circuits covered in Chapter 7 of the *Code,* and those modifications are recognized by or found in Article 250. However, Article 250 does not apply to Chapter 8 at all except where Chapter 8 references Article 250. Thus, for communications circuits, only certain portions of the circuits and equipment are required to be grounded. For example, in communications circuits, the boxes and raceways in a building are usually not required to be grounded, but the incoming conduit, cable sheath, and primary protector generally do require grounding. Other limited energy circuits and the boxes, raceways, and equipment supplied by the circuits are frequently not required to be grounded unless some specific condition such as exposure to lightning increases the hazard. The specific grounding requirements for different types of circuits and their associated equipment are covered in Chapters 3 through 6 of this book.

WIRING METHODS

For most users of the *NEC,* the most obvious differences between ordinary circuits and limited energy circuits are in the wiring methods employed. Wiring methods used for limited energy circuits need not meet the same standards as methods used for ordinary power and lighting circuits. Because shock and fire hazards are significantly reduced due to the energy limits, protection of the circuits from damage and protection of people from the circuits is not as critical as it is in circuits used for power and lighting. Each of the types of limited energy circuit modifies the general rules of Article 300 and Chapter 3 of the *NEC* in important ways. The requirements for boxes, derating of conductors, physical protection, burial depths, and other issues are eliminated or significantly modified for many power-limited circuits. These modifications and the specific wiring methods that are required or permitted for various limited energy systems are discussed in Chapters 3 through 6 of this book.

SUMMARY

This book is intended to help a user of the *NEC* understand and apply the *Code* to limited energy circuits. In order to properly interpret *Code* requirements, a user must understand the general organization of the *NEC*. Limited energy circuits may be installed under modifications to the general rules of the *Code* or, in the case of the communications systems covered in its Chapter 8, may not be subject to the general requirements of the *NEC*. How the rules for limited energy circuits differ from the general rules is one of the primary subjects of this book.

CHAPTER **2**

Low Voltage Versus Limited Energy

DEFINITIONS

The term *low voltage* is used often but in many different ways, depending on the application. Low voltage may mean 10 volts or less (Article 517); 15 volts or less (Article 680); 24 volts or less (Articles 551 and 552); 30 volts AC or less or 60 volts DC or less (Article 620); or 600 volts or less (Articles 110 and 490). So-called low voltage lighting is increasingly popular and appears as a category in displays and catalogs. Much of this lighting operates at 12 volts AC. Yet the *Code* rules for such lighting are found in Article 411, Lighting Systems Operating at 30 Volts or Less, and that article does not use the term low voltage. The actual meaning of low voltage depends on the context and is not really useful except in a specific context.

As mentioned, in contexts in which significantly higher voltage levels predominate, some standards use the term low voltage to refer to 600 volts or less. In the *NEC*, low voltage means 600 volts or less only when discussing systems that are over 600 volts. In much the same way, many equipment catalogs refer to switchboards and panelboards rated 600 volts or less as "low voltage switchgear."

These same catalogs may refer to equipment rated 15,000 volts or so as "medium voltage." In such contexts, "high voltage" refers to equipment rated 45,000 volts or perhaps 100,000 volts or more. In comparison, for most users of the *NEC*, "high voltage" means over 600 volts or over 1,000 volts, and "medium voltage" has no meaning.

OVERVIEW OF LOW VOLTAGE APPLICATIONS

The following summaries describe the characteristics of some systems covered by the *NEC* that may be referred to as low voltage, or that may provide similar functions. However, as noted in the summaries, these systems are not all necessarily limited energy systems. Although some of these systems use special wiring methods, the permitted methods are selected to be compatible with and provide adequate protection for the circuits described. Some systems and wiring methods are limited to specific occupancies to reduce the exposure of unqualified persons to the systems. These descriptions are not comprehensive and are intended only to provide a basis for comparison.

Lighting Systems Operating at 30 Volts or Less (Article 411)

- Power is limited to a maximum 30 volts and 25 amperes for the secondary circuits, and a maximum 20 amperes for the supply circuits.
- Listed systems are required. The power supplies cannot be field-constructed from other components.
- Wiring methods must be listed for the use or ordinary *NEC* Chapter 3 methods must be used. Some special cables are available, but most such cables are primarily for outside use. See Figure 2.1 for an example of a low voltage lighting system.

FIGURE 2.1 Low voltage under-cabinet lighting.

Intrinsically Safe Systems (Article 504)

- These systems have intrinsic energy limits that prevent a spark from producing enough energy to ignite a specific type of flammable mixture. Although these systems may not present a shock hazard, reducing the shock hazard is not their primary purpose, and they are not automatically considered to be Class 2 or Class 3 circuits. They are intended for application in hazardous (classified) locations.

- Ordinary or special wiring methods may be used, including methods intended for the limited energy circuits covered by Chapters 7 and 8 of the *NEC*.

- Separation requirements are more stringent than for other types of limited energy circuits as two intrinsically safe circuits are required to be separated in some manner, even from each other.

- Nonincendive circuits are similar to intrinsically safe systems in terms of permitted wiring methods, but their design and permitted applications are different. See Sections 500.7(F), 500.7(G), and 500.7(H).

Audio Signal Processing, Amplification, and Reproduction Equipment **(Article 640)**

- This article anticipates a wide range of power levels. Typical output voltage ratings are 25, 70.7, and 100 volts.
- Output wiring is generally treated as Class 1, Class 2, or Class 3 and wiring methods can be chosen from Article 725. Class 1 systems are not considered to be limited energy systems.
- Special "technical power" systems are permitted in restricted areas of nonresidential occupancies. These systems are covered by Article 647.

Sensitive Electronic Equipment **(Article 647)**

- These systems are often referred to as "technical power" and operate at 120 volts line-to-line, and 60 volts to ground (see Figure 2.2).
- Ordinary wiring methods and devices are used, but the systems are restricted to limited areas and occupancies that are closely supervised by qualified persons.
- Some specific requirements are more restrictive than for other systems. For example, voltage drop is limited to 1.5 percent on branch circuits. For ordinary branch circuits the *NEC* includes no such requirements, although some recommendations are provided.

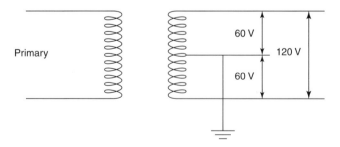

FIGURE 2.2 Technical power system.

Circuits and Equipment Operating at Less Than 50 Volts (Article 720)

- Covers only systems not specifically covered by Articles 411 (low voltage lighting; see earlier summary), 551 (recreational vehicles and parks), 650 (pipe organs), 669 (electroplating), 690 (solar photovoltaic systems), 725 (remote control and signaling; see Chapter 3 of the book), and 760 (fire alarm systems; see Chapter 4 of this book).
- Ordinary wiring methods are used, but conductors must be minimum 12 AWG copper in general, and minimum 10 AWG copper if supplying more than one appliance.

Class 1, Class 2, and Class 3 Remote-Control, Signaling, and Power-Limited Circuits (Article 725)

- These systems are distinguished by use and power supply.
- Class 1 circuits must be wired using ordinary methods, but Class 2 and Class 3 circuits may use special methods. Smaller conductor sizes are permitted with ordinary wiring methods than would be permitted for branch circuits.
- Generally, these circuits must be separated from other types of circuits, although Class 2 and Class 3 circuits may be intermixed with other limited energy circuits, and Class 1 circuits may be intermixed with certain functionally related power or lighting circuits.
- These are among the most common of the limited energy circuits. See Chapter 3 herein for more information.

Instrumentation Tray Cable: Type ITC (Article 727)

- These systems are limited to industrial locations where service is by qualified persons only.
- Type ITC (instrumentation tray cable) cable may be used only

with instrumentation and control circuits. These circuits would usually be Class 1 circuits and would usually require ordinary wiring methods, but this article provides alternative wiring methods in industrial occupancies. Although the circuit uses and energy levels suggest Class 1 circuits, the circuits covered by Article 727 are not considered to be Class 1 circuits and are not subject to the requirements for Class 1 circuits.

- Voltage is limited to 150 volts and the current is limited to 5 amperes. The power supplies are not necessarily inherently power-limited. The 5 ampere limitation may be met through the use of overcurrent devices, usually supplementary fuses.

Fire Alarm Systems (Article 760)

- These systems may be classified as power-limited fire alarm (PLFA) or nonpower-limited fire alarm (NPLFA) depending on the power supply. Some systems may use both types of circuits and power supplies.
- NPLFA systems and circuits are limited to 600 volts.
- PLFA systems must have power supplies that are listed as power-limited.
- Wiring methods, installation, and circuit separations are similar to Class 2 and Class 3 systems.
- Either ordinary methods or special cable types may be used. Smaller conductor sizes are permitted with ordinary wiring methods than would be permitted for branch circuits.

Optical Fiber Cables and Raceways (Article 770)

- These systems are inherently nonelectrical and the fibers are nonconductive; however, the cables may be classified as conductive, nonconductive, or composite, based on the presence or absence of metallic strength elements or electrical conductors.
- Special optical fiber raceways are permitted.

- Cables installed inside buildings are required to be listed.
- Optical fiber cables may be mixed with electric circuits in the same cable or raceway if they are functionally associated.

Closed-Loop and Programed Power Distribution (Article 780)

- This type of system generally uses a hybrid cable that includes conductors for power, communications, and signaling, all under one jacket. Originally these systems were intended primarily for use in home automation systems. Figure 2.3 illustrates a typical layout for so-called smart house wiring using

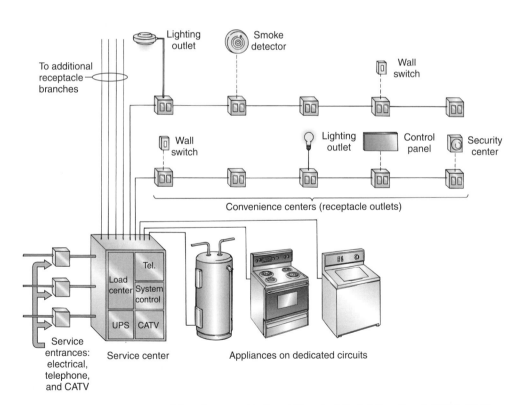

FIGURE 2.3   "Smart house" wiring. (Source: *National Electrical Code® Handbook*, NFPA, 2002, Exhibit 780.1)

Article 780 wiring methods. This technology has not been widely used.

- Signaling circuits are limited to 24 volts.
- Signaling circuits are monitored and outlets must be deenergized for certain fault conditions, including a loss of signal. In effect, only those outlets in use are energized.

Communications Circuits (Article 800)

- The *NEC* does not specify voltage or power limitations for communications circuits. Industry standards and product listing standards dictate power levels in these circuits.
- Specifically listed and marked cables are required.
- Grounding and intersystem bonding is required for most cables or cable shields, entrance raceways, and primary protectors.
- Clearances from conductors of other systems are specified and must be maintained.

Radio and Television Equipment (Article 810)

- Clearances from conductors of other systems must be maintained.
- Grounding and intersystem bonding is required for metal structures supporting antennas and for antenna discharge units.
- Special wiring methods such as coaxial cables are permitted, but ordinary wiring methods are permitted for some uses.

Community Antenna Television and Radio Distribution Systems (Article 820)

- Coaxial cable may be used with energy-limited sources only. Such sources are limited to 60 volts or less.

- Clearance from conductors of other systems must be maintained.
- Grounding and intersystem bonding is required for coaxial cable shields.
- Cables installed within buildings must be listed.

Network-Powered Broadband Communications Systems (Article 830)

- A signal from the network can be separated by a network interface unit into separate signals for voice, audio, video, and data (interactive services) as illustrated by Figure 2.4.
- Low power systems are limited to 100 volts and 250 volt-amperes.
- Medium power systems are limited to 150 volts and 250 volt-amperes. (High power systems are not recognized.)

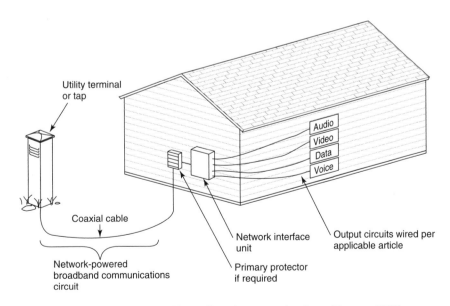

FIGURE 2.4 Network-powered broadband communications. (Source: *NEC®* *Changes 1999*, NFPA, 1999, p. 455)

- Specifically listed and marked cables are required.
- Grounding and intersystem bonding is required for most cables, the network interface unit, and primary protectors.
- Clearances from conductors of other systems are specified and must be maintained.

From the preceding descriptions, it should be evident that so-called low voltage circuits have their energy limited to various degrees. The installation requirements and permitted wiring methods for such circuits also vary. Primarily those applications considered to be power limited are covered in Chapters 3 through 7.

SUMMARY

The term low voltage is frequently used but not well defined. Many users of the term confuse it with limited energy, or low risk, and then assume that the ordinary requirements of the *Code* do not apply. Users of the *NEC* should understand that the level of risk depends on several variables, the voltage being only one factor. Limited energy or power-limited circuits are treated differently from other circuits under the *NEC* because both voltage and current limitations are applied. These limits provide a measure of protection from shock and fire hazards in most applications, but in some applications only the shock hazard or only the fire hazard is reduced. Because limited energy circuits may present a hazard under some conditions, even the special rules for these circuits are sometimes further modified for certain conditions. Many articles in the *Code* cover circuits that may be considered to be low voltage or limited energy or both. The actual rules that apply in any specific case depend on the energy limitations, the use of the circuits, the occupancy, and the exposure of persons to the circuit in question.

CHAPTER

Remote-Control and Signaling Circuits

Article 725 of the *National Electrical Code®* (*NEC®*) covers circuits that are distinguished from other types of circuits, especially power and lighting circuits, by use and power limitations. Remote control and signaling are covered by Article 725, and by their nature, most remote-control and signaling circuits require little energy to operate. Signals can be carried with very low power because the signals either directly power a low-energy device such as a doorbell, or the signals are effectively amplified through the use of relays or equivalent solid state devices. Of course, the actual power limits depend on the specific application, so Article 725 sets up classifications of circuits that are based on power levels. The lower the power, the more deviation Article 725 permits from normal *Code* rules. For example, in higher power Class 1 circuits, ordinary wiring methods are required, but smaller wire sizes are permitted. The lowest power Class 2 circuits can use not only smaller wire sizes, but entirely different wiring methods, and other normal requirements are also significantly modified. In this chapter, we examine the classifications of these circuits, look at how the normal wiring rules are modified, and see what additional restrictions are placed on these circuits and their wiring. We also see how these circuits compare to circuits for computer data transmission, intrinsically safe circuits, and nonincendive circuits.

CIRCUIT CLASSIFICATIONS

The circuits covered by Article 725 are classified as Class 1, Class 2, or Class 3. These classes are different from other circuits in the *Code* in both type of use and power limitations. All three classes have similar uses, that is, remote control and signaling, which is the primary subject of Article 725. The class designations are based on power limitations: Class 1 is least limited, Class 2 is most limited, and Class 3 is between Class 1 and Class 2.

Class 1 Circuits

Class 1 circuits are subdivided into two types: Class 1 power-limited circuits and Class 1 remote-control and signaling circuits. Class 1 power-limited circuits are those that are supplied by a power supply with a rated output that does not exceed 30 volts and 1,000 voltamperes. Class 1 remote-control and signaling circuits are limited only to 600 volts with no other power limitations.

Although the names of the two types of Class 1 circuits seem to imply different uses, the uses of both types of circuits may be the same. In fact, even though Article 725 distinguishes between the two types of Class 1 circuits, the rules for their installation are identical. However, a subtle point is made in Section 725-21 that a Class 1 power-limited circuit is not restricted to specific uses, but a Class 1 remote-control and signaling circuit is restricted to remote control and signaling. For example, given the lack of a restriction on the use of a Class 1 power-limited circuit, such a circuit could perhaps be used for lighting. However, in that case, Article 411 would also apply, and the power source and other parts would have to be listed for the lighting application. In practice, most Class 1 circuits, whether power limited or not, are actually used in remote-control and signaling applications. Furthermore, since Article 725 does not offer different rules for the installation of the two types of Class 1 circuits, many *Code* users may find no advantage or practical reason for the distinction.

Some installers and designers of remote-control and signaling circuits do mistake so-called low voltage Class 1 circuits for Class 2 circuits. Most readily available Class 2, 24 volt transformers are rated at about 40 or 50 volt-amperes. For a control system that requires more than 50 volt-amperes, a larger transformer, perhaps a machine tool transformer of 500 volt-amperes or so with a 24 volt secondary, may be purchased. The (incorrect) assumption, then, is that the usual Class 2 thermostat wire or bell wire may be used with the derived circuits. However, the circuits are not Class 2 circuits just because the voltage is in accordance with other Class 2 circuits. As pointed out in Chapter 2, the current in a Class 2 circuit must also be inherently limited by the power source. Because the transformer is not a listed Class 2 power supply, the circuits are not Class 2 circuits; they are Class 1 circuits, and ordinary wiring methods must be employed.

Class 2 Circuits

Class 2 circuits are defined by their power supplies. Section 725.2 defines a Class 2 circuit as "The portion of the wiring system between the load side of a Class 2 power source and the connected equipment," as illustrated in Figure 3.1. This definition goes on to say that "Due to its power limitations, a Class 2 circuit considers safety from a fire initiation standpoint and provides acceptable protection from electric shock." In other words, the power limitations of a Class 2 circuit reduce the risk of fire or shock from the circuit to normally acceptable levels, but to be certain these power limits will be maintained, a listed Class 2 power supply must be used.

Based on the definition of a Class 2 circuit, the identification of Class 2 circuits is easy. We must simply examine the power source. In most cases the power supply is either a transformer or a similar power source, and we only have to find the appropriate listing mark to verify the classification of the source. In some cases, the identification may be indirect. For example, a garage door operator or some

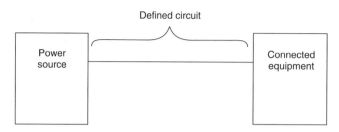

FIGURE 3.1 Limited energy circuit definition.

heating equipment may be listed as an appliance, but at the terminals for the remote controls, or on some other included information, it will likely be labeled as "*NEC* Class 2" or the equivalent, as shown in Figure 3.2. Computer systems (if they are listed) are also typically listed as "IT" (information technology) or "ITE" (information technology equipment), and if so listed, the interconnecting data circuits are considered to be Class 2.

The identification and testing of power supplies is covered in more detail later in this chapter under "Application of Class 2 and Class 3 Circuits."

FIGURE 3.2 Class 2 markings.

Class 3 Circuits

As for Class 2 circuits, Class 3 circuits are also identified and de-fined by their power supplies as shown in Figure 3.1. The definition of Class 3 circuits can be found in Section 725.2. This definition ex-plains the difference between Class 2 and Class 3 as follows: "Due to its power limitations, a Class 3 circuit considers safety from a fire initiation standpoint. Since higher levels of voltage and current than Class 2 are permitted, additional safeguards are specified to pro-vide protection from an electric shock hazard that could be en-countered." Class 3 circuits still limit power to levels that will not ordinarily initiate fires, but they do operate at higher voltages that are more likely to present a shock hazard.

Class 3 circuits are not as common as Class 2 circuits. Most mod-ern homes in the United States use multiple Class 2 circuits, but few have any permanently installed Class 3 circuits. One exception ap-pearing more frequently is in home theater and sound systems, some of which may have Class 3 audio outputs. Class 3 circuits have been fairly common in commercial sound and public address systems, such as in the system shown in Figure 3.3. These higher powered

FIGURE 3.3 Class 3 audio circuit operating at approximately 70 volts.

audio systems are finding their way into some homes as well. Class 3 circuits or their equivalent may also be found in some central fire and security systems. Again, like the Class 2 circuits, the circuits can be readily identified by examining the power supplies or the labeling of output terminals.

GENERAL REQUIREMENTS OF ARTICLE 725

As stated in Chapter 1 of this book, the articles in Chapters 5, 6, and 7 of the *NEC* can modify the requirements of its first four chapters. While Article 725 uses this authority in Section 725.3 to exclude Class 1, Class 2, and Class 3 circuits from the requirements of Article 300, in its next few subsections, it brings certain requirements back into play. Three sections of Article 300 specifically apply to all types of circuits covered by Article 725: Section 300.17 requires compliance with conduit fill limitations; Section 300.21 prohibits wiring installations from increasing the spread of fire or smoke; and Section 300.22 limits the wiring that can be installed in ducts, plenums, and other space for environmental air.

Nevertheless, the bulk of Article 300 still does not apply, at least not to Class 2 and Class 3 circuits. It may help to realize that Article 725 does reinstate the requirements of the *Code's* Chapter 3 for Class 1 wiring (see the discussion in the next section, "Application of Class 1 Circuits"), but retains the exemptions for Class 2 and Class 3 wiring. This approach makes sense because the energy levels permitted for Class 1 circuits allow these circuits be potential fire and shock hazards if not properly installed and protected. However, Class 2 and Class 3 circuits are inherently safer because the power limitations imposed on these circuits mean neither is likely to start a fire, and Class 2 circuits are not normally shock hazards either.

What types of rules have actually been eliminated for Class 2 and Class 3 circuits? Article 300 covers many general requirements for wiring methods, most of which are intended to protect circuit conductors from damage, reduce the likelihood that wiring can start a fire, and limit the exposure of people to the wiring. So we find rules in Article 300 that govern the installation and grouping of conductors to limit circuit impedances and reduce the effects of induction on raceways and enclosures. These rules are not as important with low energy circuits because smaller currents mean smaller induced currents. Minimum burial depths are another issue in Article 300. Again, since the limited energy circuits are not much of a hazard, accidentally digging into a Class 2 or Class 3 circuit is not very hazardous from a shock or fire initiation standpoint. Certainly the function of a circuit may be lost, but the loss of function is usually not a primary concern of the *NEC*, unless loss of function itself creates a hazard.

Another significant elimination is the rule in Section 300.15 that requires a box at splice, pull, and junction points. Such boxes are not required for Class 2 and Class 3 circuits, and the terminations and splices may be made in the open as shown in Figure 3.3. However, the requirements for boxes are reinstated for Class 1 circuits. A look through Article 300 will give the reader a more complete idea about the nature of the many rules that apply generally, but that do not apply to Class 2 and Class 3 circuits.

A caution or warning is appropriate at this point: Just because a rule in Article 300 does not apply does not mean that another rule directed at the same issue may not be provided in Article 725. A good example of this situation can be found in Section 300.11, which is concerned with securing and supporting of wiring methods. Section 300.11(A) prohibits the use of a ceiling grid system for the support of cables or raceways. Sections 300.11(B) and 300.11(C) prohibit raceways and cables from being used to support other wiring or other equipment. Section 300.11(B) permits certain such uses for some raceways, including the permission to attach a Class 2 circuit to the outside of a raceway where both are related to the same equipment, as illustrated in Figure 3.4.

It could be argued that the prohibited use of raceways as support for Class 2 circuits does not apply to Class 2 circuits because Article 300 does not apply to Class 2 circuits. While that may be true, Article 300 still has the authority to prohibit the use of a power system raceway for any other purpose. Article 300 does not cover plumbing or mechanical installations either, but it can prohibit the use of a raceway to support pneumatic tubing, for example.

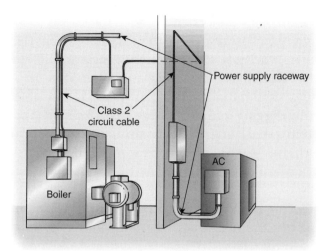

FIGURE 3.4 Raceway used as support for a related Class 2 circuit. (Source: *National Electrical Code® Handbook,* **NFPA, 2002, Exhibit 300.11)**

However, support of cables is also covered in Article 725. Section 725.58 says "Class 2 or Class 3 circuit conductors shall not be strapped, taped, or attached by any means to the exterior of any conduit or other raceway as a means of support." This section then goes on to recognize the special rule for Class 2 circuits and their related power conductors illustrated in Figure 3.4 and described in Section 300.11(B).

As previously mentioned, Section 300.11(A) prohibits the use of a ceiling grid to support raceways or cables. On its face, this rule does not seem to apply to Class 2 or Class 3 circuits. However, Section 725.6 requires that Class 1, Class 2, and Class 3 cables and conductors that are installed exposed on the outer surface of ceilings and sidewalls be supported by structural components of the building. Therefore, even though there is no direct reference to Section 300.11, the installation requirements are essentially the same because a suspended ceiling grid is not a structural component of the building. Of course, this requirement only applies to exposed cables and conductors. Section 725.6 does refer to Section 300.4(D) for protection of cables run parallel to framing members, but other concealed cables and conductors are covered only by the requirement that they be installed in a "neat and workmanlike manner."

In addition to the restriction on the use of a grid ceiling for support, Section 725.5 prohibits an accumulation of cables from preventing access to equipment above a grid ceiling or other suspended ceilings. Figure 3.5 illustrates the misuse of a grid ceiling in that the cables are supported by the grid and have accumulated to the point of interfering with the removal of a ceiling panel that is intended to provide access to equipment.

Two other requirements that apply generally to Class 1 as well as to Class 2 and Class 3 circuits are found in Sections 725.9 and 725.10. Section 725.9 simply applies the requirements of Article 250 to the circuits and equipment supplied by Class 1, Class 2, and Class 3 circuits. While this section may seem to apply the general rules of Article 250 to the circuits covered by Article 725, Article 250 actually provides special rules for remote-control and signaling circuits. We discuss this application more thoroughly when we cover specific Class 2 and Class 3 circuit requirements and applications later in this chapter. For now, the application can be summarized as shown in Figure 3.6 by noting that Article 250 requires enclosures and raceways of remote-control and signaling circuits to be grounded if the system itself is required to be grounded, that is,

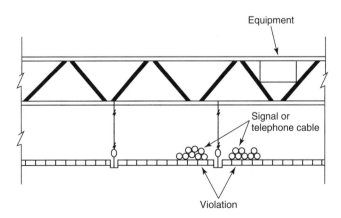

FIGURE 3.5 Accumulation of cables supported by a grid ceiling.

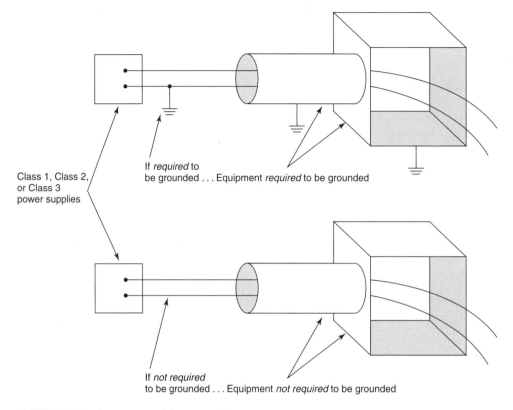

Class 1, Class 2, or Class 3 power supplies

If *required* to be grounded . . . Equipment *required* to be grounded

If *not required* to be grounded . . . Equipment *not required* to be grounded

FIGURE 3.6 Equipment requiring grounding.

if the system is required to have a grounded conductor. On the one hand, many, perhaps most, Class 2 circuits are not required to be grounded, so any boxes, raceways, or other enclosures for those conductors are not required to be grounded. On the other hand, many Class 1 circuits are required to be grounded. For example, 120 volt control circuits are very common. This voltage system is required to be grounded by Section 250.20(B). Since the system is required to be grounded (have a grounded circuit conductor), any metallic enclosures and raceways that house the 120 volt Class 1 circuits must also be grounded.

Section 725.10 requires all Class 1, Class 2, and Class 3 circuits to be identified at terminal and junction points. This identification is to be done "in a manner that prevents unintentional interference with other circuits during testing and servicing." In other words, remote-control and signaling circuits must be marked, labeled, or tagged somehow so that the circuits can be readily recognized and not confused with other conductors and circuits. The differentiation is needed not only to distinguish remote-control and signaling circuits from power and lighting circuits, but also to distinguish between different classes of remote-control and signaling circuits. This is especially important where, for example, Class 2 circuits have been wired using Class 1 wiring methods or reclassified as Class 1 circuits. In either case, ordinary wiring methods would be used rather than the special cable types usually employed with Class 2 circuits, so the wiring method alone is not a reliable way to identify the circuit classifications.

In some cases, Class 2 or Class 3 circuits are *required* to be reclassified and installed as Class 1 circuits. Consequently the use of an ordinary wiring method of a type that provides physical protection for the circuit conductors is required. This requirement for reclassification is found in Section 725.8. Reclassification is required in which a remote-control circuit is used for safety control equipment and in which the failure of the circuit would result in a "direct fire or life hazard." A common example of such a circuit might be found in a boiler control circuit as shown in Figure 3.7. Consider that the high-pressure limit on a steam boiler keeps the boiler from reaching a dangerous condition. The failure of the boiler control circuit could create a significant hazard if the circuit were shorted to simulate a closed limit switch and the boiler was not shut down in an overpressure condition. Such a circuit failure would be the direct cause of the hazard, so the circuit would have to be classified as Class 1 and installed in conduit or otherwise provided with physical protection. In contrast, a circuit used to control room temperature could fail either open or closed, and although the result could certainly be an inconvenience, the discomfort would not itself be a life or fire hazard. In Figure 3.7, the critical controls are 120 volts, so they would be classified as Class 1 anyway, and no reclassification is required, but the 24 volt thermostat could remain as a Class 2 circuit.

Another example that is often used to distinguish between a hazard caused by a circuit failure and one incidental to a circuit failure

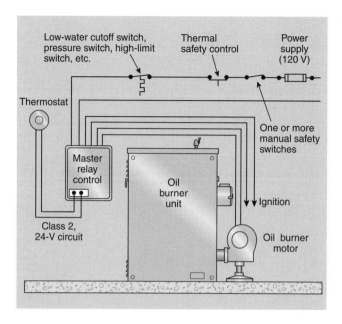

FIGURE 3.7 Automatic oil burner unit and controls. (Source: *National Electrical Code® Handbook*, NFPA, 2002)

is a nurse call system in a hospital. Patient room bathrooms are often supplied with a nurse call cord or pull switch. If someone were incapacitated in such a bathroom and the nurse call system failed due to damage to the signaling circuit, it could be a significant problem. However, the hazard would be directly due to whatever caused the incapacitation. The hazard in this case is incidental to the circuit failure, and not caused directly by the failure, so the nurse call circuit can be wired as a Class 2 circuit, as many are.

APPLICATION OF CLASS 1 CIRCUITS

Class 1 remote-control and signaling circuits are very common, especially in commercial and industrial settings where many motors are employed. Frequently, the power for a motor control circuit is tapped directly from the branch circuit supplying the motor. A transformer or fuses or both may also be installed to create or protect this circuit. These circuits are commonly 120 volts or more. By definition, they are Class 1 circuits, although Article 430 may provide less restrictive requirements for overcurrent protection of these conductors than would be allowed under the Article 725 requirements. Section 725.3(F) recognizes these special rules. Article 430 (along with Section 310.5) also allows smaller conductor sizes for control circuits than are permitted for branch-circuit conductors. Where the motor control circuits are not tapped from the motor branch-circuit protective devices, the rules of Article 725 apply. Figure 3.8 shows a motor controller in which the control circuits are tapped directly from the controller terminals that will be on the load side of the motor branch-circuit short-circuit and ground-fault protective device.

How does Article 725 modify the requirements for Class 1 circuits? According to the Fine Print Note found under Section 725.1, Article 725 modifies the requirements of Chapters 1 through 4 of the *NEC* "with regard to minimum wire sizes, derating factors, overcurrent protection, insulation requirements, and wiring methods and materials." In the following sections of this chapter, we first

FIGURE 3.8 Tapped motor control circuit.

examine each of these issues as they relate to Class 1 circuits. We also review grounding requirements for Class 1 circuits. Then we examine the same issues as they relate to Class 2 and Class 3 circuits.

Minimum Wire Sizes

The minimum size of conductors is given in Section 310.5. For branch circuits, the minimum size is 14 AWG copper. However, Section 310.5 provides 10 exceptions for specific uses and specific equipment. One of the exceptions is for Class 1, Class 2, and Class 3 circuits. Of course, Article 725 can modify the requirements of Article 310 without "permission" from Article 310.

Section 725.27 provides the special rules for sizing of Class 1 circuit conductors that permit the use of 16 AWG and 18 AWG copper. Although this section does not specify copper, Section 110.5 says that "Where the conductor material is not specified, the material and sizes given in this *Code* shall apply to copper conductors." The use of 16 AWG and 18 AWG sizes is acceptable as long as the loads are limited to the ampacities given in Section 402.5. The ampacities for larger wires are given in Section 310.15. Most of the types of wires listed in Article 310 are not available in sizes smaller than 14 AWG. Flexible cords may also be used in accordance with Article 400. Note that the ampacity given in Table 402.5 limits the load on the wires, so for an 18 AWG conductor the load is limited to 6 amperes, and for a 16 AWG the load is limited to 8 amperes.

Article 402 covers fixture wires. Normally, fixture wires are permitted to be used only as part of luminaires (lighting fixtures) or similar equipment, or for connecting to luminaires. Fixture wires are not permitted as branch-circuit conductors. However, Section 725.27(B) modifies these restrictions somewhat by permitting certain types of fixture wires to be used for Class 1 circuit conductors. Types other than fixture wire are also permitted if listed for Class 1 use. Two types of fixture wire that are often used for Class 1 circuits are Type TFN and Type TFFN. Table 402.3 describes these wire types. They are essentially equivalent to Type THHN as both are thermoplastic insulations, usually with nylon jackets. Type TFFN is only a flexible stranded conductor (probably 19-stranded), while Type TFN may be either solid or 7-stranded. The fact that Type TFFN is a flexible stranded conductor should not be taken to mean that the wire is intended for applications in which flexibility is required in normal use. According to Section 402.10, fixture wire is not intended for use where it is subject to bending and twisting in use. If such use is anticipated, a flexible cord is probably a better choice.

Derating Factors

"Derating factors" generally refers to the temperature correction factors that appear at the bottom of the allowable ampacity tables, and to the adjustment factors for more than three conductors in a raceway or cable that appear in Section 310.15(B)(2)(a). Section 725.28 says that the derating factors of Section 310.15(B)(2)(a) do not apply to Class 1 circuit conductors if those conductors are loaded to 10 percent or less of their ampacities, even when there are more than three such current-carrying conductors in a cable or raceway. The application of derating factors to power conductors is not changed by this rule except where Class 1 conductors are mixed with power conductors in the same raceway or cable, the Class 1 conductors are not counted as current-carrying conductors unless they carry more than 10 percent of their rated ampacities. Figure 3.9 illustrates an application in which derating factors do not apply. Conduit fill limits apply to both power and Class 1 circuits.

Note that the only derating factors actually mentioned in Section 725.28 are the "adjustment factors" and not the "correction factors" used for temperature adjustment. Ambient temperatures and operating temperatures of associated conductors must still be considered. Section 402.5 states that "No conductor shall be used under such conditions that its operating temperature exceeds the temperature specified in Table 402.3 for the type of insulation involved." A Fine Print Note then references Section 310.10 that provides further information on the temperature limitations of conductors.

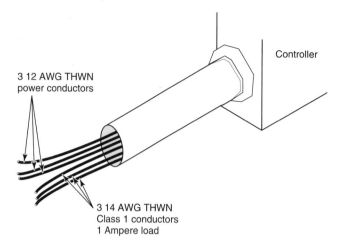

3 12 AWG THWN
power conductors

Controller

3 14 AWG THWN
Class 1 conductors
1 Ampere load

FIGURE 3.9 Derating not required.

Section 402.5 does not say how temperature issues should be considered or how ampacities should be adjusted. However, Table 402.3 does provide temperature ratings for conductor types. The types mentioned earlier, TFN and TFFN, are both 90°C conductors. Although the *NEC* never specifically says so, the correction factors from the 90°C column of Table 310.16 or Table 310.17 (the factors are the same) should also be usable for temperature corrections for TFN or TFFN fixture wires.

Another, perhaps more significant, temperature consideration in selecting Class 1 conductors is that conductors cannot be associated together in such a way that the operating temperature of any conductor is exceeded. Section 402.5 and Section 310.10 both refer to this issue. For example, suppose we select from Section 725.27(B) a Type TF or TFF for our Class 1 circuit conductors. Figure 3.10 illustrates this example. From Table 402.3 we find that these are both 60°C conductor types. If we were to mix these conductors with Type THHN conductors in a raceway as permitted by Sections 725.26 and 725.28, the THHN conductors could not be used at their 90°C ampacities, or even at the lower 75°C ampacities. The THHN conductors would be limited to 60°C ampacities to avoid overheating of the 60°C Class 1 conductors.

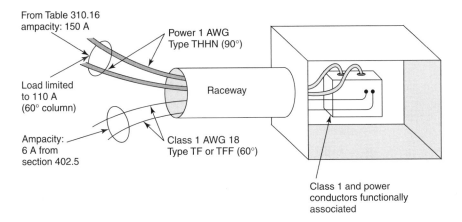

From Table 310.16
ampacity: 150 A

Power 1 AWG
Type THHN (90°)

Load limited
to 110 A
(60° column)

Raceway

Ampacity:
6 A from
section 402.5

Class 1 AWG 18
Type TF or TFF (60°)

Class 1 and power
conductors functionally
associated

FIGURE 3.10 Conductors with lower temperature ratings.

Overcurrent Protection

Section 725.23 covers overcurrent protection for Class 1 circuit conductors. This section says the conductor ampacity may be used and
the derating factors of Section 310.15 may be disregarded in calcu-
lating an ampacity. This rule applies only to conductors 14 AWG or
larger. Overcurrent protection for smaller conductors is specified to
not exceed 7 amperes for 18 AWG and 10 amperes for 16 AWG. Thus,
for all conductors under this rule, the maximum values for overcurrent protection are or may be somewhat higher than the ampacities of the wires. This provision should not be surprising if the
nature of the loads on Class 1 conductors is considered.

Most remote-control and signaling conductors are not heavily
loaded and in many cases are loaded only intermittently. The most
likely overcurrents are short circuits and ground faults. In other
areas of the *NEC,* where overloading is a significant risk, overcurrent protection is usually not permitted to exceed the ampacity of a
conductor. In control circuits, overloading is not likely in most cases.
In fact, the Exception to Section 725.23 recognizes other rules of the
Code that permit or require other levels of overcurrent protection.

Consider the rule for motor control circuits. In motor control
circuits as covered in Section 430.72, overcurrent protection is often

permitted to be provided by the motor branch-circuit short-circuit and ground-fault protective device. The values of overcurrent protection permitted in Table 430.72(B) apply to three basic conditions: (1) where separate protection is provided, usually in the form of a supplementary fuse (Column A); (2) where separate protection is not provided and the conductors do not leave the motor control equipment enclosure (Column B); and (3) where separate protection is not provided and the conductors do leave the motor control equipment enclosure (Column C). If the motor branch-circuit short-circuit and ground-fault protective device exceeds the values in Column B or Column C as applicable, then separate protection must be provided under Column A. The effect of these rules is that where separate protection is provided, it may as well protect the conductor at its ampacity, but otherwise only short-circuit and ground-fault protection is required for the motor control circuit. As pointed out previously, overloading is not likely on these types of circuits anyway.

Even the rule for motor control circuits has exceptions. One concerns the situation in which the opening of a control circuit overcurrent device could create an increased hazard. In such a case, only short-circuit and ground-fault protection is required, and that protection may be provided by the motor circuit device without restriction on its rating. Fire pump circuits are an obvious and frequently mentioned example, but a crane control circuit is another example. Specifically, Section 610.53 generally permits protection at 300 percent of the control circuit conductor ampacity. This section also permits a hot metal crane control circuit to be protected by the motor branch-circuit device without regard to its rating.

In general, overcurrent protection for Class 1 circuits should be located where the conductors receive their supply. However, Section 725.24 permits alternative locations in much the same manner as the tap rules of Section 240.21, and the rules for Class 1 conductors differ a bit from the ordinary feeder tap rules. Section 725.24(B) permits a Class 1 conductor to be tapped from a feeder where the overcurrent device for the larger conductor is sized to also protect the tap. Section 725.24(C) permits Class 1 conductors that are at least 14 AWG to be tapped on the load side of a controlled light and power circuit as long as the overcurrent device does not exceed 300 percent of the conductor ampacity. Using this rule, a 14 AWG conductor could be tapped from up to a 50 ampere circuit that feeds a group of high-intensity discharge (HID) parking lights, as illustrated in the example in Figure 3.11, which shows only the control wiring

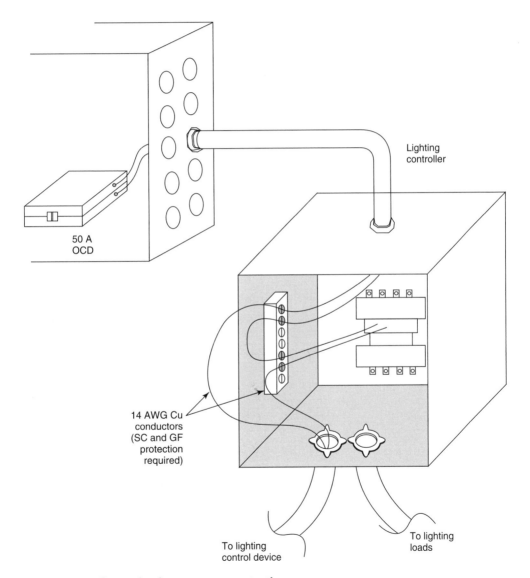

FIGURE 3.11 Class 1 circuit overcurrent protection.

and not the power wiring. The conductor could then be run to a time clock or some other control device to direct the operation of the lighting contactor or similar device that is used to control the lights. This rule is very similar, but not identical, to the rules for tapped motor control circuits found in Section 430.72. It could be argued that higher rated devices could be used, based on the ampacity of a 14 AWG wire. However, according to Section 210.23, 50 ampere circuits are the largest rated circuits that are permitted for lighting loads circuits with more than one outlet. Higher ratings could be used in motor circuits.

Sections 725.24(D) and (E) permit Class 1 conductors to be protected by the overcurrent device on the supply side of a single-phase transformer with a two-wire secondary or a similar listed Class 1 electronic power supply. In both of these rules, the ampacity of the Class 1 conductors is multiplied by the secondary-to-primary or output-to-input voltage ratio, and the result must not be less than the rating of the primary overcurrent device.

Figure 3.12 illustrates an example of a Class 1 electronic power supply where the input fuse is permitted to be used to protect the output conductors. The ampacity of the 16 AWG output conductors is 8 amperes from Section 402.5. The output-to-input voltage ratio is 0.25, so the maximum rating for an input fuse that can protect the

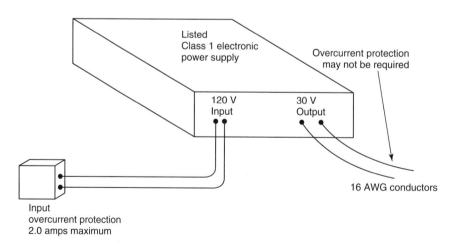

FIGURE 3.12 Class 1 circuit. Protection using primary overcurrent device.

Class 1 conductors is $8 \times 0.25 = 2$ amperes. If the power supply were a transformer, the primary fuse would also have to be selected to protect the transformer in accordance with Section 450.3(B) or as permitted for control power transformers in Section 430.72(C).

Insulation Requirements

Insulation requirements are not significantly altered for Class 1 circuits. Section 725.27(B) requires Class 1 circuit conductors to be suitable for 600 volts. The thermoset and thermoplastic insulations listed in Table 310.13 and permitted for branch circuits or feeders are all 600 volt or higher rated insulation types. Although Table 402.3 does include some 300 volt rated insulation for some fixture wires, none of the 300 volt types are listed in Section 725.27(B) as suitable for Class 1 wiring. Note that although Class 1 circuits may be classified as power limited, there are no special wiring methods for these circuits. Even those Class 1 circuits that are limited to 30 volts are required to be wired with 600 volt insulated conductors.

Wiring Methods and Materials

As discussed, Class 1 circuits are treated differently from branch circuits with regard to minimum wire sizes and types of wire, the use of derating factors, and overcurrent protection. Earlier in this chapter it was stated that Article 725 excludes remote-control and signaling circuits from the requirements of Article 300, but that certain provisions of Article 300 are reinstated, such as the conduit fill limits of Section 300.17 and the requirements for wiring in ducts, plenums, and other air-handling spaces.

For Class 1 circuits, Section 725.25 applies all of Article 300 along with whatever articles in Chapter 3 of the *Code* are applicable to the particular wiring method chosen. Interestingly, Section 725.3 says "Only those sections of Article 300 referenced in this article shall apply to Class 1, Class 2, and Class 3 circuits." Section 725.25 does not reference any sections of Article 300; it references the entire article. When this specific but broad reference to Article 300 was added in the 2002 *NEC,* Code Making Panel 16 said "It is the intent of the panel that all sections of Article 300 apply to installations of Class 1 circuits." This means, for example, that Section 300.15, which requires boxes at splice, junction, outlet, and pull points, will apply to Class 1 circuits. Nevertheless, this reinstatement of all sections of Article 300 is limited by Article 300 itself, since for example, Part II of Article 300 applies only to installations over 600 volts and Class 1 circuits are limited to 600 volts. Prior to the 2002 *Code,* Section 725.25 referred only to "the applicable articles from Chapter 3," which many authorities took to include Article 300. The intent is now clearer.

Basically, there are no special wiring methods for Class 1 wiring. Users of Class 1 circuits can choose from the ordinary wiring methods of Chapter 3 of the *Code.* The general rules of Chapter 3 are modified as previously explained, primarily with regard to the size and protection of conductors, but otherwise Class 1 wiring is treated pretty much the same as branch-circuit wiring. This treatment includes any occupancy restrictions, so if a particular wiring method cannot be used in an assembly occupancy or a hazardous location for a branch circuit, that wiring method generally cannot be used for a Class 1 circuit either. One explicit example of this situation is found in Section 725.29. This section requires a Class 1 circuit that extends overhead and outside of a building to meet the requirements of Article 225, which, in turn, covers outside branch circuits and feeders. Since Section 725.29 only applies to aerial extensions from a building, we would focus primarily on those rules of Article 225 that apply to overhead conductors, such as conductor clearances from ground.

Circuit Separations

This is not to say that there are no special rules for Class 1 wiring. Although we may choose from the same methods used for branch circuits, Article 725 imposes some additional restrictions on installations of Class 1 circuit conductors. Probably the most obvious restriction concerns the mixing of Class 1 circuits with other circuits in raceways and enclosures. These rules, found in Section 725.26, are modifications of the general rules found in Article 300. In Article 300, conductors of different circuits may share a common enclosure if they are all 600 volts or less, and all conductors are insulated for the maximum circuit voltage in the enclosure. Generally, conductors of circuits that are over 600 volts cannot be mixed with circuits that are 600 volts or less. Article 725 permits mixing of Class 1 circuits with power circuits only when they are functionally associated. This is true even though all the conductors are likely rated 600 volts. Figure 3.13 shows an example of a motor circuit and a Class 1 circuit that are functionally associated and share a common raceway and enclosure.

Section 725.26(A) allows two or more Class 1 circuits to occupy the same raceway, cable, or enclosure as long as they are all insulated for the maximum voltage of any circuit contained. Since, as we have seen previously, Class 1 wiring is required to have 600 volt insulation, and Class 1 power supplies are limited to 600 volts, this requirement is met almost automatically if the other requirements for Class 1 circuits are met.

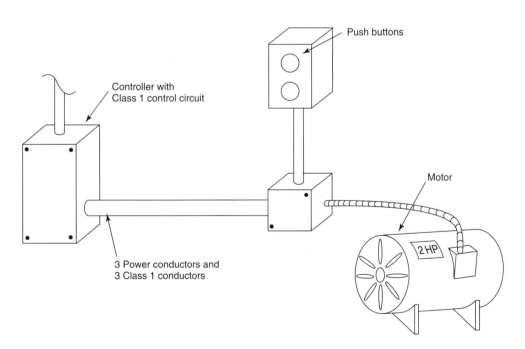

FIGURE 3.13 Functional association.

Section 725.26(B) permits Class 1 conductors to share a raceway, cable, or enclosure with power circuits only if they are functionally associated, are run to the same control center, or are separated. In manholes, the separation can be provided by enclosing the Class 1 conductors in Type UF cable or metal-enclosed cables, or by providing a fixed separation by fixed nonconductors or fastening to supports. In a cable tray the separation can be provided by a fixed barrier or by installing either the power conductors or the Class 1 conductors in a metal-enclosed cable.

Grounding of Class 1 Circuits and Equipment

The general application of Article 250 to remote-control and signaling circuits are reviewed under the previous section in this chapter, "General Requirements of Article 725." In that discussion we note that enclosures and raceways of remote-control and signaling circuits are required to be grounded only where the *system* is required to be grounded. A grounded system is a system in which one of the normal circuit conductors is grounded. A grounded conductor is a conductor that has been intentionally connected to earth or a body that serves in place of earth. (These terms are defined in Article 100.)

Section 250.112(I) says that equipment related to remote-control and signaling circuits has to be grounded if the system is required to be grounded according to Part II or Part VII of Article 250. Part II covers alternating current systems and Part VII covers direct current systems. Specifically, Section 250.20 addresses AC systems, and Section 250.162 covers DC systems, although other sections or exceptions may modify these requirements. If and only if a *system* is required to be grounded under Part II or Part VII, the *equipment*, that is, the metallic raceways, boxes, fittings, and enclosures for the circuit conductors, is also required to be grounded.

The criteria for determining whether a Class 1 system must be grounded include the voltage and phase relationships of a system; a consideration of the ways a system could be grounded; the maximum voltage to ground; the system from which the Class 1 system is derived; and the conditions of use of the system, such as whether the

Class 1 system runs outside a building as overhead conductors. A complete discussion of the grounding of systems is outside the scope of this book, but three examples to illustrate how grounding requirements apply to Class 1 circuits follow.

Example 1

A 120 volt AC, single-phase, two-wire Class 1 control circuit is derived from a 480 volt system through a 500 volt-ampere transformer that is located inside a control panel as shown in Figure 3.14. If either leg of the secondary of this transformer were grounded, the voltage to ground would be 120 volts. This condition matches the conditions of Section 250.20(B)(1), so the system must be grounded and Sections 250.112(I) and 250.86 would require the associated metallic equipment, including raceways and enclosures, to be grounded. Because the transformer is smaller than 1,000 volt-amperes, Section 250.30(A)(1), Exception No. 2 and Section 250.30(A)(2), Exception permit the bonding jumper conductor to be a 14 AWG conductor, and the grounded transformer frame or enclosure to be used as the electrode for this separately derived system.

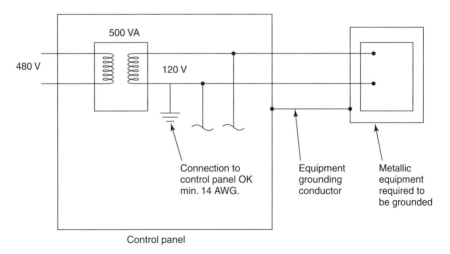

Control panel

FIGURE 3.14 First example of required grounding of Class 1 circuits.

Example 2

A 24 volt AC two-wire control circuit is derived from a 1,000 volt-ampere transformer supplied from a three-phase 480 volt system as shown in Figure 3.15. The exact nature of the 480 volt system is unknown by the control system designer, but the voltage to ground of a 480 volt system could be 277 volts if the system is 480/277 volts, four-wire grounded wye; or 480 volts if the system is 480 volt, three-wire grounded delta; or unknown if the system is an ungrounded 480 volt system. (The voltage to ground of a 480 volt undergrounded delta system is *defined* as 480 volts.) In any case, the system is either ungrounded, or the voltage to ground exceeds 150 volts. Therefore, Section 250.20(A) requires the 24 volt system and related equipment to be grounded. Again, as in the first example, the bonding jumper could be a 14 AWG conductor, and the grounded transformer enclosure or frame could serve as the electrode.

Example 3

This example is illustrated by Figure 3.16. A 24 volt AC, two-wire control circuit is derived from a 250 volt-ampere transformer supplied from a three-phase 480 volt system. However, the control system in this case also uses some 120 volt AC Class 1 circuits, so the 24 volt AC system is derived from the 120 volt system, which, in turn, is derived from the 480 volt system. In other words, the 120 volt system is derived in the same manner as in Example 1, and the

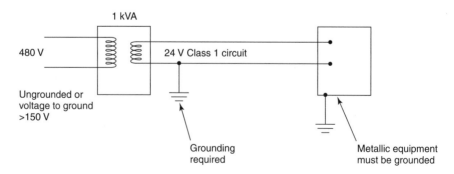

FIGURE 3.15 Second example of required grounding of Class 1 circuits.

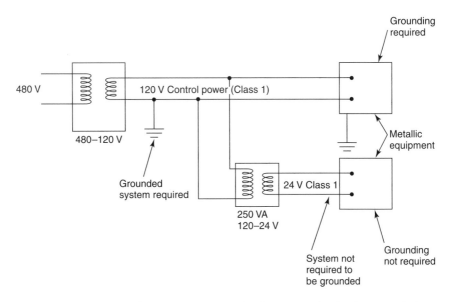

FIGURE 3.16 Example of grounding Class 1 circuits not required to be grounded.

120 volt system and equipment are required to be grounded. If the 120 volt transformer is over 1,000 volt-amperes, it will have to be treated as an ordinary separately derived system, otherwise the exceptions noted in Example 1 may be applied. However, from the standpoint of the 24 volt system, it is now derived from a system that is required to be grounded and where the voltage to ground is less than 150 volts. So unless the 24 volt system runs outside the building and overhead, Section 250.20(A) does not require the 24 volt system to be grounded. Since the system is not required to be grounded, the equipment supplied solely by the 24 volt system is also not required to be grounded.

Note that schemes similar to Example 3 are often used in manufactured equipment, and sometimes the manufacturers ground the 24 volt systems anyway. The fact that the system is grounded does not trigger the requirement for the equipment to be grounded. The equipment is required to be grounded only when the system is *required* to be grounded according to Section 250.112(I).

A caution to readers: The grounding requirements of Section 250.20 are grouped into Sections 250.20(A), 250.20(B), and 250.20(C) based on system voltages, or what is defined in Article 100 as the "voltage of a circuit." However, within Sections 250.20(A) and 250.20(B), the Sections most likely to apply to Class 1 circuits, the requirements key on the "voltage to ground," which is also defined in Article 100. For example, in a typical 240/120 volt system as shown in Figure 3.17, the voltage of a circuit can be 240 volts, but the voltage to ground will be 120 volts. Therefore, in such a system, even if a 24 volt transformer has a 240 volt primary, the voltage to ground for that primary is still 120 volts. The *NEC* is precise in this sort of language and must be read with the same precision.

A second caution: The requirements for grounding in hazardous locations are much more stringent. In these areas grounding and bonding is required for all metal raceways regardless of system voltage. See Section 250.100 and the discussion later in this chapter under the heading "Related Circuits—Intrinsically Safe and Nonincendive Circuits."

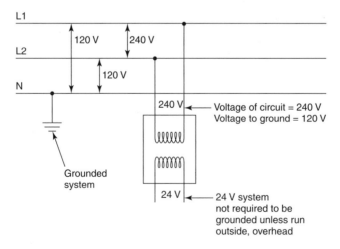

FIGURE 3.17 Voltages in a circuit.

APPLICATION OF CLASS 2 AND CLASS 3 CIRCUITS

Class 2 circuits appear in many common applications in all types of occupancies. Class 2 circuits are routinely used in temperature controls for homes and businesses, doorbells, door openers, lighting controls, security systems, irrigation controls, communications accessories such as caller identification boxes, battery chargers for cell phones, power supplies for computer accessories, and many other applications. (Power supplies for some computer accessories may not be marked as Class 2 but may instead be marked as "information technology equipment.") Computer interconnecting wiring for data communications is another rapidly expanding use of Class 2 circuits. Class 3 circuits are less common, but not rare. Some other devices use systems that are not precisely Class 2 or Class 3 systems but use either Class 2 or Class 3 wiring methods. Audio systems are good examples of such systems. However, the power levels in audio systems vary significantly, so some speaker systems may be treated as Class 1 circuits.

We have seen that Class 2 and Class 3 circuits are defined by their power supplies. The power supplies limit the total energy in a Class 2 or Class 3 circuit to a defined maximum value that will not be exceeded even with a short circuit on the load side of the power supply. These inherent energy limitations are what make Class 2 and Class 3 circuits limited energy or power-limited circuits. The actual permitted values of voltage and current that are permitted to be delivered by Class 2 and Class 3 circuits are found in Chapter 9 of the *Code* in Tables 11(A) and 11(B). Table 11(A) gives the power source limitations for AC systems, and Table 11(B) gives the limitations for DC systems. According to the notes for the tables, the tables are provided in the *Code* "for listing purposes." They should not be confused with design criteria for field-constructed power supplies. Class 2 and Class 3 power supplies must be listed and marked to indicate the class of supply and the electrical rating. Figure 3.18 shows a Class 2 power supply with the required markings.

Although Tables 11(A) and 11(B) provide listing information, other information and specific test procedures are included in the

FIGURE 3.18   Class 2 power supply marking.

product standards for Class 2 and Class 3 power supplies. Class 2 power supplies are tested according to Underwriters Laboratories' Standard UL 1310, *Standard for Safety for Class 2 Power Units*. Class 3 power supplies are tested according to UL 1012, *Standard for Safety for Power Units Other Than Class 2*. These product standards apply to individual power supplies and not to the power supplies that are an integral part of other equipment. Power supplies that are part of other equipment, such as audio equipment or information technology equipment, are investigated as a part of that equipment. These product standards do provide many design considerations for safe use as well as testing procedures and criteria for operation of the power supplies. Tables 11(A) and 11(B) provide only the power limitations and none of the other requirements for construction and testing of power supplies.

Power supplies may take different forms. One of the more common forms is a direct plug-in unit such as those used for powering irrigation sprinkler time clocks, household security systems, and small battery chargers. This type of Class 2 power supply is covered by UL 1310. Standard UL 1310 also covers similar power supplies that are connected by cord and plug, such as power supplies for some computer accessories and toys. As previously noted, power supplies for Class 3 circuits are covered by UL 1012. Standard UL 1012 also covers power supplies that are not listed as Class 3 or limited to Class 3 power levels, such as battery chargers for wheelchairs and the like. Many Class 2 and Class 3 circuits are supplied by transformers that are intended for a field connection to a source of supply. These transformers may be equipped with mounting means for attachment directly to a box cover, to an enclosure knockout, or to a mounting panel in a cabinet. Such transformers are covered by UL 1585, *Standard for Class 2 and Class 3 Transformers*. Since transformers are inherently AC devices, UL 1585 includes the power output limits included in Tables 11(A) for AC systems, but also includes many other detailed specifications and tests required of listed Class 2 and Class 3 transformers. Power supplies cannot be considered to be equivalent to listed power supplies just because they apparently meet the power limitations of Tables 11(A) or 11(B).

The notes to Tables 11(A) and 11(B) provide some important requirements that may not be immediately obvious from the marking of a power supply. One sentence in the table notes says "Power sources designed for interconnection shall be listed for the purpose." Interconnecting power sources may be desired for reliability or redundancy, but when power sources are connected in parallel (see Figure 3.19), the total output is changed and the power limitations for the circuit class may be defeated. Therefore, the units must be listed for interconnection. This requirement also appears in Section 725.41(B).

Another note to the tables indicates that Class 2 power sources may become Class 3 power sources when used in wet locations. Markings to this effect are common on Class 2 transformers, but not always recognized. The UL-required marking is "CLASS 2 NOT WET, CLASS 3 WET," as shown in Figure 3.20. This marking is required when the output terminals are designed for field connections and the AC voltage exceeds 21.2 volts peak (about 15 volts rms), which is half the 42.4 volts peak normally permitted at the output of a power supply for the voltage not to be considered a risk of electric

FIGURE 3.19 Parallel Class 2 power supplies.

shock. The point is that Class 2 power supplies are supposed to limit voltages to a level that is reasonably safe from the standpoint of shock, but that level may be changed in a wet location. Class 3 power supplies do not consider safety from the standpoint of shock.

For most users of Class 2 and Class 3 power supplies, the types of insulations, physical dimensions, dielectric withstand tests, and other specific requirements for listing are of little or no interest. Most users and inspectors only need to know how to identify a listed power source. Class 2 power supplies are required by UL standards to be marked "Class 2 Battery Charger," "Class 2 Transformer," "Class 2 Power Supply," or "Class 2 Power Unit." Class 3 markings are similar. However, as noted in Section 725.41(A)(4) and the notes to Tables 11(A) and 11(B), listed information technology computer equipment is not required to be marked as Class 2 or Class 3. The equipment is only required to be marked as listed information technology equipment, usually abbreviated "Info. Tech. Equip." or "ITE," as shown in Figure 3.21.

FIGURE 3.20 Wet location marking.

According to UL 1950, *Safety of Information Technology Equipment, Including Electrical Business Equipment,* output terminals on listed information technology equipment that are provided for field wiring from Class 2 power supplies must be marked with the voltage rating and "NEC Class 2" or "NEC Class 2 Output." The marking is to be adjacent to the terminals. Output connectors are assumed to be supplied from a limited power source unless they are marked otherwise, or unless the installation instructions say otherwise. For output connectors that are not from a limited power source, the marking or installation instructions must identify the circuit type, the intended cable type, or the circuit voltage rating.

FIGURE 3.21 Information technology equipment marking.

Section 725.41(A)(4) states that "Listed information technology (computer) equipment limited power circuits" are considered to be Class 2 or Class 3 sources. Standard UL 1950 defines the power levels that are permitted for a limited power source. The output voltage is limited to 30 volts AC or 60 volts DC and 100 volt-amperes for inherently limited sources or 250 volt-amperes for sources that require the use of an overcurrent device. Compared to the power levels permitted for Class 2 and Class 3 power sources in Tables 11(A) and 11(B), the computer limited power sources are actually restricted to a subset of the values permitted for Class 2 power sources. Because of this categorization, information technology equipment limited power circuits are generally considered to be Class 2 circuits under Article 725.

We have established the fact that the classification of a Class 2 or Class 3 circuit depends on the power supply. Generally the classification of the power supply can be readily determined from the marking on the power supply, or, if the power supply is part of other listed equipment, from the marking on that equipment or at the limited energy terminals of the equipment. Once the classification of the power supply has been determined and verified, how are the circuits to be treated? The Fine Print Note to Section 725.1 says that Article 725 modifies the requirements of Chapters 1 through 4 of the *Code* with regard to five issues: minimum wire sizes, derating factors, overcurrent protection, insulation requirements, and wiring methods and materials. As we did for Class 1 circuits, we now examine each of these issues, as well as the issue of grounding, as they relate to Class 2 and Class 3 circuits.

Minimum Wire Sizes

Installers of Class 2 or Class 3 circuits have three choices of wiring methods:

1. The circuits can be (or may be required to be) reclassified as Class 1 circuits and installed as Class 1 circuits, or may be wired using Class 1 methods without reclassifying the circuit.
2. The circuits can be installed using cable types specifically listed for the circuit type.
3. The circuits can be installed using substitute listed cable types.

If the circuits are reclassified or otherwise use Class 1 methods, the minimum wire size will be the same as for Class 1 circuits, that is, 18 AWG. Otherwise, the *NEC* specifies listed cables, but does not specify a minimum size of conductor for those cables. Where single conductors are used in Class 3 circuits, the minimum size is specified as 18 AWG. However, most Class 2 and Class 3 wiring is done using the listed cable types specified by the *NEC*, and for those cables the minimum conductor sizes are dictated by the product standard.

The product standard for Class 2 and Class 3 cables is UL 13, *Standard for Safety for Power-Limited Circuit Cables*. According to this standard, the smallest copper conductor in Class 2 cable is 30 AWG, and the smallest copper conductor in Class 3 cable is 24 AWG. Smaller conductors are actually permitted in Class 2 and Class 3 cable if the tested breaking strength of the cable is at least 25.0 pounds force (111 Newtons) or equivalent. Generally, Class 2 and Class 3 cables are most readily available with conductors in sizes between 24 AWG and 12 AWG. Type PLTC (power-limited tray cable) is not permitted by the *NEC* to have conductors smaller than 22 AWG or larger than 12 AWG. Larger sizes are available in other Class 2 and Class 3 cables for special uses such as for audio applications and for the central member in a coaxial cable. Typically the size selected is based on performance requirements or manufacturer's specifications. For a given application, a manufacturer of a control system or device will frequently specify a minimum size for up to some maximum circuit length, with larger sizes specified for longer lengths. Overcurrent protection ratings are not used to size Class 2 and Class 3 conductors because the power supplies are inherently power-limited, and the *NEC* does not provide ampacities for conductors smaller than 18 AWG.

Section 725.51 also establishes requirements for the wiring on the supply side of a Class 2 or Class 3 power source. This section requires that the overcurrent protection ahead of the power source be no greater than 20 amperes. However, the exception to this section permits the input leads of a transformer to be smaller than 14 AWG if the leads are not longer than 12 inches. The minimum size for these transformer leads is 18 AWG and the conductor must be of a type permitted for Class 1 circuits. The exception does not change the 20 ampere maximum overcurrent rating, it just permits transformer leads to be protected at a value higher than their ampacity. This permission is very similar to what is permitted for listed appliances where the load is known and overloading is not a significant risk. For example, Sections 240.4(A) and 240.4(B) permit 18 AWG cord on a listed appliance or 18 AWG fixture wire to be supplied from a 20 ampere branch circuit.

Derating Factors

Article 725 does not say anything about derating of Class 2 and Class 3 circuit conductors installed as Class 2 or Class 3 circuits. Nor does Article 310 provide ampacities for the conductor sizes most commonly used for these circuits. This omission is not an oversight. There is really no need for such rules since the power supplies are inherently energy limited even under short-circuit conditions. Article 725 permits the use of small conductors because there is little risk of overloading, power supplies are limited, and the levels of power limitation keep the circuits from being fire hazards even if the conductors were to become overloaded, damaged, or otherwise faulted.

When the term *derating factors* is used with regard to the conductor ampacities of Article 310, it refers to "correction factors" applied to higher than normal ambient temperatures as well as "adjustment factors" applied to more than three current-carrying conductors. The low current levels that are typical in Class 2 and Class 3 circuits make the conductors similar in heating characteristics to Class 1 conductors that are similarly loaded. Class 1 conductors that carry loads not over 10 percent of their ampacities are not required to be counted as current-carrying conductors. Such conductors are simply not adding significant heat to the installation. Nevertheless, all conductors and cables must be applied within their ratings.

Class 2 and Class 3 cables do have temperature ratings. These temperature ratings are required to be marked on the cables if they are other than 60°C. Many cables have no temperature markings, and where there are no markings, and the cable is listed, the cable can be assumed to be rated for 60°C. Cables with higher temperature ratings are common, especially up to 90°C, and will be marked as such. Depending on the properties of the insulation and jacket materials, the temperature rating of a Class 2 or Class 3 cable may be as high as 250°C. Section 110.3(B) says that listed equipment must be installed in accordance with the listing and labeling, and this rule applies to the cable types that are listed for use with Class 2 and Class 3 circuits.

Overcurrent Protection

Overcurrent protection is inherent in the power supplies for Class 2 and Class 3 circuits. The only significant rule about separate overcurrent protection for these circuits applies solely to the supply side of a power supply. The overcurrent device on the supply side is limited to 20 amperes. This rule, found in Section 725.51, is also mentioned in the previous discussion of minimum wire sizes.

Insulation Requirements

The required insulation ratings for Class 2 and Class 3 cables are found in Section 725.71(F). This section requires Class 2 cables to have insulation rated at not less than 150 volts. Class 3 cables must have insulation rated for at least 300 volts. These values are significantly less than the 600 volt insulation required for Class 1 circuits and for branch circuits and feeders. However, unlike Class 1 wires, Class 2 and Class 3 cables are not permitted to have the voltage rating marked on the cables. Cables that have listings in addition to the Class 2 or Class 3 listings are permitted to have voltage markings if one of the additional listings requires the markings. For example, Type PLTC cable is required to be marked with its 300 volt rating. This marking may help distinguish Type PLTC cable from Type TC (tray cable), which typically has a 600 volt rating, and Type ITC (instrumentation tray cable), which has 300 volt insulation but is not marked. Voltage markings on power-limited cables may be misconstrued to imply suitability for higher power uses, such as for Class 1 circuits or power or lighting circuits.

Standard UL 13, *Standard for Safety for Power-Limited Circuit Cables,* specifies insulation thicknesses and tests for Class 2 and Class 3 cables. Type PLTC cable insulations are tested to the same standard as Class 3 cables. The standard permits a manufacturer to choose between test methods in some cases. One option for Class 2 cables that are nonintegral is a dielectric withstand test. (Nonintegral cables have separate insulated conductors under an overall jacket, whereas integral cables look like a variation of a flat extension cord or "zip cord," in which the insulation is thicker and also serves the function of a jacket.) In a dielectric withstand test, a Class 2 cable must withstand a DC voltage of 1,250 volts for at least 2 seconds with no faults. The similar test for a Class 3 or PLTC cable requires the cable to withstand 2,500 volts for at least 2 seconds with no faults. These tests are required to be performed on 100 percent of the production by the manufacturer. In addition, samples of Class 3 and PLTC cables must pass an insulation resistance test that is conducted as part of the follow-up testing of a manufacturer's products. Insulation resistance tests are not required for Class 2 cables. (These examples of testing procedures are not complete descriptions of the requirements, and are included here only for a comparison of the characteristics of Class 2 and Class 3 cables.)

Wiring Methods and Materials

As stated in the discussion of minimum wire sizes, there are three choices for wiring methods for Class 2 or Class 3 circuits. The circuits can use Class 1 wiring methods whether they are reclassified as Class 1 or not; the circuits can be installed using cable types specifically listed for the circuit type; or the circuits can be installed using substitute listed cable types.

The main rule of Section 725.52 permits either the use of Class 1 wiring methods under Section 725.52(A) or Class 2 or Class 3 wiring methods under Section 725.52(B). Section 725.52(A) permits the use of Class 1 wiring methods without mention of reclassifying the circuit. Such circuits would have to continue to be clearly marked as a Class 2 or Class 3 circuit in accordance with Section 725.42, and continue to be subject to the separation requirements for Class 2 and Class 3 circuits. If a Class 2 or Class 3 circuit is installed entirely or partially using Class 1 wiring methods, it remains a Class 2 or Class 3 circuit. However, if a Class 2 or Class 3 circuit is required to be *reclassified* by Section 725.8, or if the circuit is reclassified for some other reason, it becomes a Class 1 circuit. Circuits that are reclassified become Class 1 circuits even though they continue to be supplied by a Class 2 or Class 3 power source, and thus all requirements for Class 1 circuits would apply, including overcurrent protection requirements. However, a reclassified circuit would be subject to the Class 1 separation requirements rather than the more stringent Class 2 and Class 3 separation rules.

Section 725.52 must be applied carefully. Section 725.52(A), Exception No. 2 contains permissive language as described in Section 90.5(B). While Section 725.8 requires reclassification under certain conditions, Exception No. 2 to Section 725.52(A) simply permits reclassification. Section 725.52(A) permits the use of Class 1 wiring methods without reclassifying or modifying the requirements for marking and circuit identification. Exception No. 2 permits the reclassification of the circuit, which then forces the use of Class 1 methods and requires the removal of the Class 2 or Class 3 circuit markings. There is a difference between using Class 1 wiring methods and making the circuit a Class 1 circuit, and both possibilities are recognized in Section 725.52.

Perhaps the most popular choice for Class 2 or Class 3 wiring methods comes from Section 725.52(B). This section permits the use of special wiring types that are not generally permitted for any other uses. Thus, if we have a Class 2 circuit, we can pick a Class 2 cable type that is suitable for the occupancy and location within the occupancy.

Section 725.71 describes the applications for which specific cable types are listed. The basic cable types are CL2, CL3, and PLTC. Type PLTC is a Class 3 wiring method, similar in requirements to CL3, but Type PLTC is also required to be sunlight resistant and is not required to be so marked. Type PLTC cable is intended for use in outdoor cable trays. It may also be used in some Class I and Class II, Division 2 areas and is considered to have the "gas/vapor-tight continuous sheath" referred to in Section 501.5(D) and 501.5(E). Type PLTC may be used for direct burial if marked "dir bur," "direct burial," or "for direct burial." Cables marked for direct burial must pass a crushing test and a water absorption test. Figure 3.22 shows examples of cables and their markings.

Cable types CL2 and CL3 are suitable for general use. General use does not include use in plenums, ducts, risers, or other environmental air-handling spaces unless the cables are installed in metal raceways as permitted in Section 300.22. Types CL2 and CL3 may also have additional letters to designate special uses. The letter *P* stands for "plenum"; *R* stands for "riser"; and *X* indicates limited use. Additional letters or designations may be included to indicate

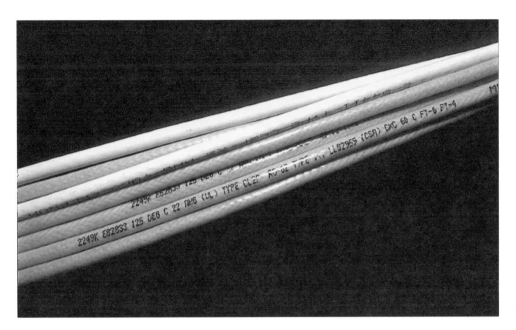

FIGURE 3.22 Cable markings.

other ratings, such as cables that are also suitable for communications (Type CM) or cables that are sunlight resistant, suitable for direct burial, or intended for other special uses such as thermocouple extension wire.

Thermocouple extension wire is also marked with a code to indicate the combination of metals used in the wires. Some cables are marked with the *NEC* article with which they are intended to comply. Cables intended for data wiring are usually marked with a category number such as "Cat 5" that indicates their performance in data transmission. As already mentioned, cables with temperature ratings other than 60°C will be marked with the temperature rating. Some of these special markings may also appear on Type PLTC.

Using the marking scheme described, we see that Type CL2P or CL3P can be used in ducts, plenums, or other space for environmental air without enclosing the cables in a raceway. The *P* suffix indicates that the cable has been tested and passes the plenum flame test that measures flame propagation and smoke density. To be suitable for a plenum application, such cables must not spread a fire rapidly or produce a lot of smoke that could be transported and distributed through a building by the air-handling system. The required markings make identification and selection of cables easy for installers and inspectors.

Section 725.3(C) does apply Section 300.22 to Class 2 and Class 3 circuits, but it also permits the use of CL2P and CL3P cables in air-handling spaces. Although Sections 725.3(C) and 725.61(A) modify Section 300.22 with respect to permitted wiring methods, Section 300.22 still restricts wiring in ducts and plenums. According to Sections 300.22(A) and 300.22(B), no wiring is permitted in ducts that handle loose stock or vapor, and only that wiring that must be in a duct for functional purposes may be in other fabricated ducts. Figure 3.23 illustrates the uses that are and are not permitted in ducts and plenums.

Cables marked CL2R or CL3R are intended for riser use. Risers are installations that extend vertically through more than one floor. The *R* suffix means that the cable passes a riser flame test. To be installed in a riser, a cable should be slow in propagating flames so that a fire will not spread from floor to floor by following the cable. In lieu of riser cable, other cables may be installed in raceways or in fire-

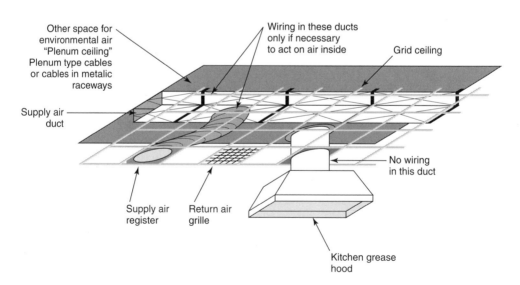

FIGURE 3.23 Wiring in ducts and plenums.

proof shafts with firestops at each floor. Risers in one- and two-family dwellings are not required to use riser cables or raceways.

Type CL2X and Type CL3X cables are intended for limited uses in dwelling units or in raceways only. These cable types are also tested for flame spread, but they are tested to a somewhat lesser standard than types CL2 or CLR. In one- and two-family dwellings these types may be used in risers. Cables with the X suffix that are also not more than ¼ inch in diameter may be installed anywhere in one- and two-family dwellings and in nonconcealed spaces in multi-family dwellings. They may be used in nonconcealed spaces in other types of occupancies in which the exposed cable is not more than 10 feet long. Types CL2X and CL3X can also be used in any occupancy if they are installed in a raceway. Due to the numerous restrictions on this type of cable, many installers choose to use a general purpose or riser rated cable for more flexibility in application. Type CL2X cable occupies the lowest position on the cable substitution hierarchy, so any of the other types of cables we have discussed in this section can be used in its place.

Many specific applications and the cable types and installation methods permitted for those applications are listed in Section 725.61. For example, only Type PLTC is intended for use in cable trays outdoors, but any of the Class 2 and Class 3 cable types except the limited use CL2X and CL3X types can be used in indoor cable trays. One special application for which there is no CL2 or CL3 cable type available is for installations under carpet. For this application, Type CMUC (communications under carpet) may be used. In fact, CM types can be used as substitutions for any of the CL2 or CL3 types, given similar applications and suffixes. The last subsection of Section 725.61 and Table 725.61 provide a list of the permitted substitutions. These substitutions are also illustrated in Figure 725.61, which is reproduced here as Figure 3.24.

An issue relating to Class 2 and Class 3 circuits that has sometimes been controversial is the application of conduit fill limits to these circuits. In previous editions of the *NEC*, Article 725 circuits were excluded from Article 300 requirements except where there was a specific reference to some section of Article 300. Article 300 and Section 300.17 that covers conduit fill were then brought back for

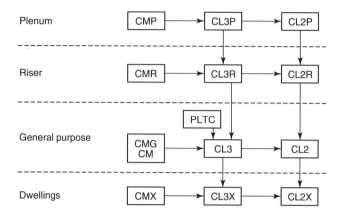

Type CM—Communications wires and cables
Type CL2 and CL3—Class 2 and Class 3 remote-control, signaling, and power-limited cables
Type PLTC—Power-limited tray cable

A → B   Cable A shall be permitted to be used in place of cable B.

FIGURE 3.24 Substitutions for Class 2 and Class 3 cables. (Source: *National Electrical Code®*, NFPA, 2002, Figure 725.61)

Class 1 circuits, but not for Class 2 and Class 3 circuits. (Some would argue this point as well.) However, specific references within the raceway articles, such as Section 358.22, still required compliance with the conduit fill limits of Table 1 in Chapter 9. The controversy centered on which reference, or lack of reference, took precedence. This controversy is resolved in the 2002 *NEC* because Section 725.3(A) now applies Section 300.17 to all circuits and equipment covered by Article 725. As a result, conduit fill limits do apply to Class 1, Class 2, and Class 3 conductors.

Circuit Separations

As for Class 1 circuits, Class 2 and Class 3 circuits are permitted to use special installation rules and wiring methods, but they are also subject to special restrictions. The primary restriction, and according to many inspectors the restriction that results in the most problems in the field, is the requirement for circuit separations. These restrictions are found in Section 725.55, which overrides the rules of Section 300.3(C)(1) where only voltage and insulation levels are considered. In Section 300.3(C)(1), any circuits under 600 volts may be mixed in a raceway or enclosure as long as the insulation on all conductors is at least equal to the highest circuit voltage. But a Fine Print Note after this rule directs the reader back to Section 725.55 for Class 2 and Class 3 circuit separations. The types of circuits that can be mixed without separations and the restrictions on these circuits are found in Section 725.56.

Class 2 and Class 3 circuits get special treatment because they are supplied from power-limited sources. These sources are carefully specified, designed, and tested to be sure that the energy limits can be relied on. To be certain that the energy limits will not be compromised, energy-limited circuits must be separated from higher powered circuits. This separation provides reasonable assurance that the limited energy circuits will remain as energy-limited circuits.

The basic separation requirement that Class 2 and Class 3 circuits will not be in close proximity to or in contact with higher energy circuits is found in Section 725.55(A). However, in Sections 725.55(B) through 725.55(J), a range of ways to maintain the separations is provided. Class 2 and Class 3 conductors may be separated by a barrier or by a raceway and still occupy the same overall enclosure with non-power-limited circuit conductors, as illustrated in Figure 3.25. Many other specific methods are listed for specific types of installations, such as installation in cable trays, in manholes, or hoistways. Generally, a physical separation of at least 2 inches is required unless separation is maintained by metal raceways, or by barriers.

In some cases, such as manholes, separation can be provided by fastening to racks or other supports, by using fixed nonconductor, or by installing the non-power-limited circuits in Type UF cable. Type NM cable is permitted in some other locations. In a cable tray, Class 2 and Class 3 conductors can be installed in Type MC cable in close proximity to power circuits, and the MC cable sheath is considered to provide the required separation. We do not discuss all the specific possibilities in this book. The reader can find a method of separation for most situations in Sections (B) through (J).

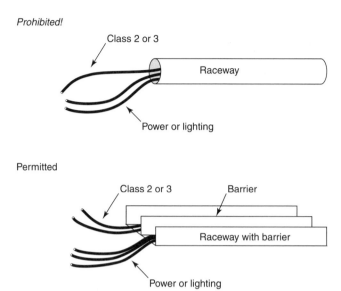

Prohibited!

Class 2 or 3

Raceway

Power or lighting

Permitted

Class 2 or 3   Barrier

Raceway with barrier

Power or lighting

FIGURE 3.25 Class 2 and Class 3 circuit separations.

Maintaining separations is a special problem in enclosures where Class 2 or Class 3 conductors must enter an enclosure to connect to the same equipment as higher powered circuits. This situation arises often in control panels and similar locations, where, for example, a Class 2 control circuit must connect to the same relay as the power circuit it controls. In such cases, two options are provided in Section 725.55(D). First, physical separations may be provided within the enclosure, but the requirement is reduced to ¼ inch from the usual 2 inches. This requirement that the conductors be "routed to maintain a minimum of 6 mm (0.25 in.) . . ." is illustrated in Figure 3.26. In order to maintain this separation, the conductors must usually be fastened in place or otherwise contained in a wire management system of some sort.

The second option is available where the circuit conductors operate at no more than 150 volts to ground. In this case, two more options are available: The conductors can be installed as Class 1 circuit conductors, or the jacket of a Class 3 or permitted substitute cable may be used for separation. Class 1 conductors are permitted to occupy the same enclosure with higher power conductors if they are

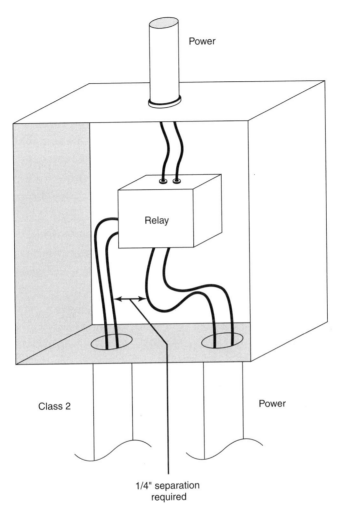

FIGURE 3.26 Separations of Class 2 and power circuits in common enclosures.

functionally associated, which they certainly are if they connect to
the same relay(s). Where Class 3 or substitute cable jackets are used
for separation, the conductors that are not contained in the cable
jacket still must meet the minimum ¼ inch spacing requirement.

When Class 2 or Class 3 conductors must enter an enclosure with
power conductors and the enclosure has only one opening, the limited energy circuits must be separated from the higher powered circuits by a nonconductor, such as flexible tubing, that is firmly fixed

in place. This requirement is found in Section 725.55(E). A common example of such an enclosure is a fan relay of the sort frequently used in heating and air conditioning applications (see Figure 3.27).

Section 725.56 lists the circuit types that may be mixed in an enclosure and the conditions under which they may be mixed. Essentially, circuits of the same type may be in the same cable, or in the same raceway or other enclosure as shown in Figure 3.28. They need not be functionally associated.

Class 2 and Class 3 circuits may be in the same cable, enclosure, or raceway if all the circuits are insulated to Class 3 requirements. Typically, this condition would mean using Class 3 or acceptable substitute cables for both the Class 2 and Class 3 circuit conductors.

Class 2 or Class 3 circuits may also be mixed in the same *cable* with communications circuits if all the circuits are classified as communications circuits, as shown in Figure 3.29. Applications similar to Figure 3.29 are very common in uses of Category 5 cables. Category 5 cable is often used for both conventional telephone (communications circuits) and computer data (Class 2 circuits), but the cable is listed as a communications or multipurpose type and all the circuits are classified as communications under the *NEC*. (See Chapter 6 in this book and Chapter 8 in the *NEC* for more information on communications cables and permissible uses and substitutions.)

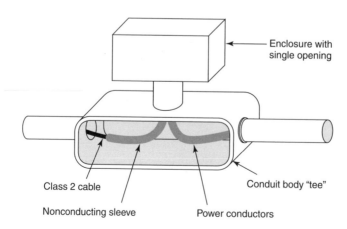

FIGURE 3.27 Separations at enclosures with a single opening.

Permitted:

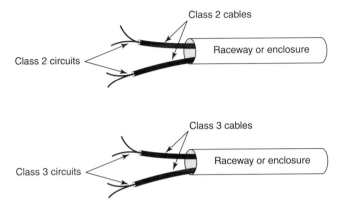

FIGURE 3.28 **Circuits of the same type in the same raceway or enclosure.**

If the Class 2 or Class 3 circuits are in their own jacketed cables, they may occupy the same raceway with certain other power-limited circuits that are also contained within their own jacketed cables, as shown in Figure 3.30. This permission does not extend to mixing jacketed cables in a raceway with non-power-limited circuits. Section 725.56(E) includes a list of the types of jacketed cables that may be mixed. Where the individual cables share a raceway or other enclosure, the jackets can serve to separate the circuits and they do not have to be reclassified. Reclassification is required only where different classes of individual conductors share a single cable under a single jacket.

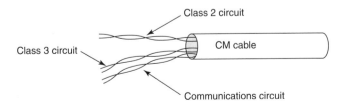

FIGURE 3.29 **Class 2 or Class 3 circuits reclassified.**

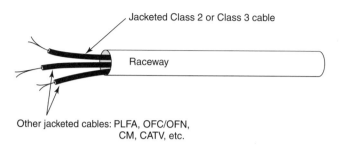

Jacketed Class 2 or Class 3 cable

Raceway

Other jacketed cables: PLFA, OFC/OFN,
CM, CATV, etc.

FIGURE 3.30 Individually jacketed cables in a raceway.

Grounding of Class 2 and Class 3 Circuits and Equipment

We have previously summarized the general application of Article 250 to remote-control and signaling circuits under the section "General Requirements of Article 725." In that section we note that enclosures and raceways of remote-control and signaling circuits are required to be grounded only where the *system* is required to be grounded. A grounded system is a system in which one of the normal circuit conductors is grounded. A grounded conductor is a conductor that has been intentionally connected to earth or a body that serves in place of earth. (These terms are defined in Article 100.)

Section 250.112(I) states that equipment related to remote-control and signaling circuits has to be grounded if the system is required to be grounded according to Part II or Part VII of Article 250. Part II covers alternating current systems and Part VII covers direct current systems. Specifically, Section 250.20 addresses AC systems, and Section 250.162 covers DC systems, although other sections or exceptions may modify these requirements. If and only if a *system* is required to be grounded under Part II or Part VII, the *equipment*—the metallic raceways, boxes, fittings, and enclosures for the circuit conductors—must be grounded.

The preceding summary of grounding requirements applies to Class 2 and Class 3 circuits in much the same way as it applies to Class 1 circuits. However, one significant difference should be noted: Class 2 and Class 3 circuits have significantly lower voltage limitations than Class 1 circuits. Even where the voltage levels are equivalent, as in 24 volt control systems, Class 2 and Class 3 systems have much lower power limitations than Class 1 circuits. Although a Class 2 system could be rated at up to 150 volts at 5 milliamperes, most Class 2 sources are less than 30 volts. Therefore, most Class 2 AC systems fall under the grounding requirements of Section 250.20(A), and under this rule, Class 2 systems and circuits are not required to be grounded unless derived from systems where the voltage to ground is over 150 volts or derived from ungrounded systems, or the Class 2 conductors run outside of buildings and overhead. If the system is not required to be grounded, the equipment, raceways, boxes, and other metallic enclosures are not required to be grounded either.

Figure 3.31 illustrates an application in which a metal box and conduit are not required to be grounded. The Class 2 system is derived from a 120 volt system that is required to be grounded and in which the voltage to ground does not exceed 150 volts. The Class 2 system is less than 50 volts and is therefore not required to be grounded. So the raceway and box are not required to be grounded. The equipment labeled "HVAC" (heating, ventilation, and air conditioning) is required to be grounded.

Class 3 circuits typically have higher voltage levels than Class 2 circuits and, therefore, are more likely to fall under Section 250.20(B).

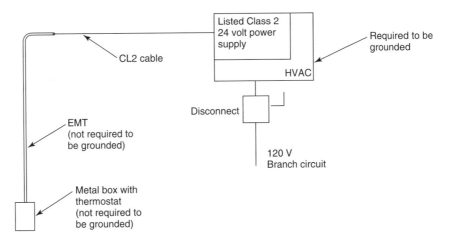

FIGURE 3.31 Equipment not required to be grounded.

They are somewhat more likely to require grounding. Nevertheless, the steps taken to determine whether a system is required to be grounded or not and, subsequently, whether the associated equipment is required to be grounded are the same for Class 1, Class 2, and Class 3 circuits.

Class 2 and Class 3 DC systems are treated somewhat differently. As pointed out previously, DC circuits are covered in Part VII of Article 250, and the grounding requirements are found in Section 250.162. Although Article 250 provides rules for AC systems under 50 volts, Section 250.162 says nothing about two-wire DC systems under 50 volts. Therefore, many common DC systems, such as 48 volt nominal systems, are not required to be grounded, and neither are the enclosures and other equipment connected solely to these circuits. Two-wire DC systems between 50 and 300 volts, as well as three-wire DC systems, are required to be grounded. There are some exceptions for the two-wire systems, but all three-wire systems, regardless of voltage, are required to be grounded.

A caution to readers: The grounding requirements of Section 250.20 are grouped into Sections 250.20(A), 250.20(B), and 250.20(C) based on system voltages, or what is defined in Article 100 as the "voltage of a circuit." However, in Sections 250.20(A) and 250.20(B), the Sections most likely to apply to Class 2 and Class 3 circuits, include the requirements key on the "voltage to ground," which is also defined in Article 100. For example, in a typical 208/120 volt system, the voltage of a circuit can be 208 or 120 volts, but the voltage to ground will be 120 volts. Therefore, in such as system, even if a 24 volt transformer has a 208 volt primary, the voltage to ground for that primary is still 120 volts. (Refer back to Figure 3.17 that illustrates the use of these terms.) The *NEC* is precise in this sort of language and must be read with the same precision.

A second caution: The requirements for grounding in hazardous locations are much more stringent. In hazardous areas, grounding and bonding is required for all metal raceways regardless of system voltage. See Section 250.100 and the discussion later in this chapter under the heading "Related Circuits—Intrinsically Safe and Nonincendive Circuits."

APPLICATIONS IN COMPUTER/DATA WIRING

As discussed previously, the circuits used for exchanging information between computers (information technology equipment) are generally classified as Class 2 circuits. However, these circuits are often reclassified as communications circuits or installed using communications cables as a substitute for CL2 cables. The broadband communications between computers now demanded by many users force the use of Category 5 or higher bandwidth copper cables, coaxial cables, or fiber-optic cables. Many coaxial cables are available as Class 2 or communications cables in addition to the CATV classification that many people associate with "coax."

The *NEC* does not treat the circuits for computer information differently from other limited energy circuits of the same classification. Some installation techniques are different or require special care to be sure that the intended data transfer rates can actually be achieved, but these issues are outside of the safety concerns of the *NEC*.

As already noted, the product standard for information technology equipment, UL 1950, requires output terminals for Class 2 circuits to be so marked. Output connectors are required to be marked (or identified in installation instructions) if they are *not* supplied by limited power sources. However, some connectors may be provided for connection to telecommunications networks, and the power supply in such circuits comes from the telecommunications network. Standard UL 1950 describes the voltage in these circuits as "TNV" (telecommunications network voltage). TNV levels are typically 48 volts DC but the voltage is permitted to be higher for a ringing signal, and the actual operating voltage has both an AC and a DC component. Since such circuits are communications circuits, they are required to use communications cables or equivalent cable types.

Standard UL 1950 permits interconnection circuits between equipment to be one of the following: SELV (safety extra low voltage), limited current circuits, TNV, or a hazardous voltage circuit. Except in special cases, ELV (extra low voltage) circuits are not permitted for equipment interconnections. SELV and ELV circuits have the same voltage limits: 42.4 volts peak or 60 volts DC. (In a sinu-soidal AC circuit 42.4 V peak is equal to 30 volts rms, the normal measured or so-called effective voltage.) However, SELV circuits must be isolated from higher voltage circuits by more than simple insulation and must maintain their voltage limitations under single-fault conditions. Limited power sources typically meet the require-ments for SELV circuits. Hazardous voltage circuits are circuits in which the voltage exceeds 42.4 peak or 60 volts DC and the cir-cuits are not classified as either TNV circuits or limited current cir-cuits. Limited current circuits may have voltages that exceed 15,000 volts, but the current is limited so that even under likely fault con-ditions, a hazardous current cannot be drawn from the circuit.

Article 645

Article 645 is titled Information Technology Equipment. Based on this title, *Code* users often assume that the article covers all such equipment. However, Article 645 is actually a permissive article that applies only if certain provisions are met. The entire article has the effect of an exception that may be used when an installation com-plies with all of the listed stipulations. Because the article is in Chap-ter 6 of the *Code*, it can modify the general rules of the first four chapters, but there is no requirement that the article be applied. Section 645.2 says the article shall apply if five conditions are met. Partly because the conditions are not required to be met, the vast majority of ITE equipment is not installed under the provisions of Article 645.

The five conditions for using Article 645 are:

1. Special disconnecting means must be provided at all princi-pal exit doors to disconnect HVAC and ITE equipment in the room.
2. HVAC systems must be dedicated to the ITE room or sepa-rated from other spaces by fire dampers.
3. Listed ITE equipment must be installed.

4. Occupancy of the room must be restricted to only those persons needed for the maintenance and operation of the ITE equipment.

5. The room must separated from other areas by fire-resistant construction.

Nowadays even large file servers and computers with processing power equivalent to what would once have been called "mainframe computers" are relatively small devices that are often installed in an equipment closet or even at an individual workstation. Large computer rooms or computer labs without restricted access are commonplace, especially in educational facilities. Raised floors are still useful for interconnecting wiring, but the underfloor space is not necessarily used to handle environmental air. Many installations simply do not need the space or cooling equipment that was once needed for similar computing power and information storage.

Nevertheless, some facilities still have a need for the type of computer room covered by Article 645. Figure 3.32 is a picture of a

FIGURE 3.32 Information technology room with raised floor.

large information technology room with a raised floor of the sort anticipated by Article 645. Facilities are free to use the article if they meet the conditions of Section 645.2. The question is, what do they get for meeting all the conditions? They get the special rules that apply under Article 645, found primarily in Section 645.5, which has five subsections [(A) through (E)] that spell out the special requirements and permissions.

Section 645.5(A) requires branch-circuit conductors to be rated at not less than 125 percent of the connected load.

Section 645.5(B) permits the IT equipment to be connected to branch circuits by cord and plug. Flexible cords up to 15 feet long with attachment plug caps or cord set assemblies may be used.

Section 645.5(C) allows separate data processing units to be connected by cables and cable assemblies that are listed for the purpose.

Section 645.5(D) allows wiring methods under a raised floor that might not be permitted otherwise, especially where the area under the raised floor is used to handle environmental air. However, according to this subsection another list of conditions must all be met:

- The floor must be of "suitable construction" and the area under the floor must be accessible. These floors usually have 2 foot by 2 foot removable tiles that rest on a structural framework. "Suitable construction" means acceptable under the building code.

- Specific wiring methods must be used. The list of acceptable wiring methods is very broad and includes nonmetallic raceways that would not be permitted by Section 300.22(C), which covers "other space for environmental air." However, Section 300.22(D) recognizes Article 645.

- Smoke detection must be provided under the raised floor so that air-handling equipment for the underfloor space will be turned off if smoke is detected. This provision is new in the 2002 *NEC*, but some local jurisdictions had previously required smoke detection and fan shutdown under other codes or prohibited the use of nonplenum rated wiring methods in the underfloor spaces.

- Openings in the raised floors must be provided with a means to minimize the entry of debris into the underfloor space. Where the openings are used for cabling, a means must also be provided to protect the cables from abrasion.

- Only certain types of cables may be used in the underfloor space. This rule has been confusing for many people, primarily because they try to read it too literally. Cables with metal jackets are permitted by the second condition and are also permitted under this rule. Type DP cables listed for the use are also permitted. The other permitted types are those complying with Section 645.5(D)(5) (a), (b), and (c). Taken literally, this permission is read to mean that the other types must comply with all three requirements, but when all three requirements are applied, the rule is nonsensical. Literally, the cables would have to be installed in a raceway; installed only with equipment manufactured before July 1, 1994; and be one of the types listed. So, taken literally, no new installations of new equipment could comply, and all cables except Types DP, MC, MI, and AC would have to be in raceways. This interpretation was not the intent. The intent was that Types MC, MI, and AC could be used, *and* Type DP could be used, *and* cables in raceways could be used, *and* older cable types listed for use with the older equipment could be used, *and* the list of (mostly) energy limited cable types could be used. In addition, the limited energy *NEC* cables and fiber-optic cables could be, but are not required to be, plenum, riser, or general purpose types as indicated by the additional letters *P*, *R*, or *G* respectively.

- Abandoned cables must be removed. This rule is also new in the 2002 *NEC*.

Figure 3.33 is a picture of a raised floor with an access panel removed showing the underfloor wiring. The photo does not necessarily show a compliant installation or clearly indicate whether all of the requirements for a raised floor have been met. For example, the tile would have to be put back in place to see the required grommets for protecting the cables coming through the floor.

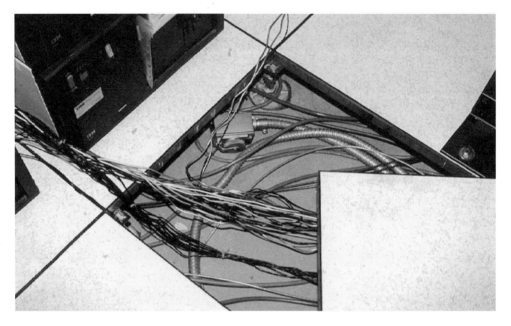

FIGURE 3.33 Accessible raised floor.

Section 645.5(E) allows wiring methods and equipment listed for use as ITE equipment or for use with ITE equipment to be installed without being secured in place. This permission allows for tremendous flexibility in the wiring of the ITE rooms that are covered by Article 645.

An important point here is that a room with a raised floor is not necessarily a room that is covered by Article 645. Nor does such a room necessarily need to be covered by Article 645. A raised floor does not necessarily have to be used to handle air, but even when the underfloor space is treated as a plenum, plenum rated wiring methods can be used. A user may still derive the flexibility and access to wiring afforded by a raised floor by meeting the ordinary requirements of the *NEC*, even in an area that does not have restricted access or that does not meet all of the other requirements of Section 645.2.

RELATED CIRCUITS—INTRINSICALLY SAFE
AND NONINCENDIVE CIRCUITS

Chapter 2 includes a brief description of the intrinsically safe circuits covered by Article 504. These circuits are a type of energy-limited circuit, and they are frequently used for purposes similar to those for which Class 2 and Class 3 circuits are used. Another type of circuit that is similar is a nonincendive circuit. Both nonincendive circuits and intrinsically safe circuits are available as methods of protection in hazardous (classified) areas. Both types of circuits are designed so that the available energy in the circuit is not sufficient to ignite a specific flammable atmosphere. Since some flammable atmospheres are more easily ignited than others, the energy levels in intrinsically safe and nonincendive circuits can vary according to the specific materials that cause an area to be classified.

Because energy levels may vary, intrinsically safe and nonincendive circuits are not automatically assumed to be Class 2 or Class 3 circuits unless the power supplies for the circuits are also listed as Class 2 or Class 3 power supplies. Class 2 and Class 3 circuits also have specific voltage limits. Intrinsically safe and nonincendive circuits could have higher voltages as long as the available current is limited to very low values. Intrinsic safety is based on energy and the characteristics of the hazardous material, not just on voltage or current. Intrinsically safe circuits must have energy limits that are maintained under both normal and abnormal conditions. Figure 3.34 is a schematic diagram of a fused zener diode barrier, one of the methods used to maintain the intrinsic safety in a circuit. The two zener diodes connected in parallel provide for voltage limitation, and one is redundant in case the other fails. The two resistors connected in series provide for current limitation, and again, one is redundant in case the other fails. The fuse adds another level of current limitation.

Both nonincendive and intrinsically safe circuits are permitted to use any of the wiring methods suitable for unclassified locations. In addition, Section 504.20 specifically permits the use of the wiring methods of Chapters 7 and 8 of the *Code* for intrinsically safe circuits. This permission is not given in Sections 501.4(B)(3), 502.4(B)(3), or

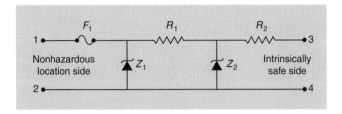

FIGURE 3.34 Fused zener diode barrier. (Source: *National Electrical Code®*, NFPA, 2002, Exhibit 504.2)

505.15(C)(1)(g) in which wiring methods for nonincendive circuits are covered. However, Section 501.4(B)(3)(3) permits multiconductor cables to be used in nonincendive circuits based only on the insulation thickness on the individual conductors. Since many CL2 cables meet or exceed this insulation thickness, those cables, but not necessarily all CL2 cables, may be used with nonincendive circuits. Also, where a particular wiring method, say, Class 2 cable, has been evaluated as part of a particular nonincendive circuit, that cable type could be used.

Chapter 5 of the *NEC* has complete authority over the wiring methods that may be used in hazardous areas. Chapter 7 cannot modify Chapter 5 in this respect. Nevertheless, Section 725.61(D)(3) also permits the use of Class 2 wiring for nonincendive circuits. Otherwise, Section 725.61(D) requires the use of Type PLTC cable for Class 2 and Class 3 circuits installed in hazardous locations unless the circuits are nonincendive for the specific atmosphere. Although this language is not absolutely clear, a conservative interpretation would require Type PLTC cable in hazardous locations, even in rigid metal conduit, unless the Class 2 or Class 3 circuit is nonconductive, is intrinsically safe, is reclassified as Class 1, or unless Class 1 wiring methods are used.

The permitted use of a specific cable type, such as CL2, does not make the circuit a Class 2 circuit. Also, even though an intrinsically safe or nonincendive circuit would appear to meet the energy limitations of Tables 11(A) or 11(B), the power source does not necessarily meet the listing requirements of Section 725.41 for Class 2 and Class 3 power sources.

Nonincendive and intrinsically safe circuits are directed at the same goal: limiting the energy in a circuit to levels below the level required to ignite a flammable atmosphere. However, intrinsically safe circuits are designed to maintain these energy limits even under abnormal conditions and therefore are permitted in Division 1 areas where the hazard is likely to exist normally or frequently. Nonincendive circuits may not maintain their energy limitations under abnormal conditions, so they are permitted only in Division 2 areas where the hazardous atmosphere does not exist normally or frequently. All nonincendive and intrinsically safe power supplies and apparatus are required to be identified for the class of location and the specific properties of the flammable or combustible materials that are present.

SUMMARY

Article 725 covers Class 1, Class 2, and Class 3 circuits used primarily for remote-control and signaling applications. These three circuit classifications covered by Article 725 are defined by the voltage and current limitations of their power sources. The effect of Article 725 is to provide special rules that modify the general requirements of the first four chapters of the *NEC*.

This chapter covers the various ways that the general rules are modified for Class 1, Class 2, and Class 3 circuits. Class 1 circuits are not limited to power levels that significantly reduce the shock or fire-starting hazards of the circuits, but Class 3 circuits do limit fire hazards, and Class 2 circuits limit both fire and shock hazards. Class 1 circuits have special rules with regard to wire sizes, derating factors, overcurrent protection, and separation from other circuits, but generally use the same insulation types and wiring methods permitted for ordinary circuits. Class 2 and Class 3 also have special rules with regard to wire sizes, derating factors, overcurrent protection, and separation from other circuits, but also are permitted to use different insulation types and wiring methods. Finally, this chapter covers the ways that the circuits of Article 725 compare to three types of related circuits: circuits for computer data transmission, intrinsically safe circuits, and nonincendive circuits.

Information Technology Equipment

645.1 Scope. This article covers equipment, power-supply wiring, equipment interconnecting wiring, and grounding of information technology equipment and systems, including terminal units, in an information technology equipment room.

FPN: For further information, see NFPA 75-1999, *Standard for the Protection of Electronic Computer/Data Processing Equipment*.

645.2 Special Requirements for Information Technology Equipment Room. This article shall apply, provided all the following conditions are met:

(1) Disconnecting means complying with 645.10 are provided.
(2) A separate heating/ventilating/air-conditioning (HVAC) system is provided that is dedicated for information technology equipment use and is separated from other areas of occupancy. Any HVAC system that serves other occupancies shall be permitted to also serve the information technology equipment room if fire/smoke dampers are provided at the point of penetration of the room boundary. Such dampers shall operate on activation of smoke detectors and also by operation of the disconnecting means required by 645.10.

FPN: For further information, see NFPA 75-1999, *Standard for the Protection of Electronic Computer/Data Processing Equipment*, Chapter 8, 8-1, 8-1.1, 8-1.2, and 8-1.3.

(3) Listed information technology equipment is installed.
(4) The room is occupied only by those personnel needed for the maintenance and functional operation of the installed information technology equipment.
(5) The room is separated from other occupancies by fire-resistant-rated walls, floors, and ceilings with protected openings.

FPN: For further information on room construction requirements, see NFPA 75-1999, *Standard for the Protection of Electronic Computer/ Data Processing Equipment*, Chapter 3.

Source: NFPA 70, *National Electrical Code*®, NFPA, Quincy, MA, 2002 edition.

645.5 Supply Circuits and Interconnecting Cables.

(A) Branch-Circuit Conductors. The branch-circuit conductors supplying one or more units of a data processing system shall have an ampacity not less than 125 percent of the total connected load.

(B) Cord-and-Plug Connections. The data processing system shall be permitted to be connected to a branch circuit by any of the following means listed for the purpose:

(1) Flexible cord and attachment plug cap not to exceed 4.5 m (15 ft).
(2) Cord set assembly. Where run on the surface of the floor, they shall be protected against physical damage.

(C) Interconnecting Cables. Separate data processing units shall be permitted to be interconnected by means of cables and cable assemblies listed for the purpose. Where run on the surface of the floor, they shall be protected against physical damage.

(D) Under Raised Floors. Power cables, communications cables, connecting cables, interconnecting cables, and receptacles associated with the information technology equipment shall be permitted under a raised floor, provided the following conditions are met.

(1) The raised floor is of suitable construction, and the area under the floor is accessible.
(2) The branch-circuit supply conductors to receptacles or field-wired equipment are in rigid metal conduit, rigid nonmetallic conduit, intermediate metal conduit, electrical metallic tubing, electrical nonmetallic tubing, metal wireway, nonmetallic wireway, surface metal raceway with metal cover, nonmetallic surface raceway, flexible metal conduit, liquidtight flexible metal conduit, or liquidtight flexible nonmetallic conduit, Type MI cable, Type MC cable, or Type AC cable. These supply conductors shall be installed in accordance with the requirements of 300.11.
(3) Ventilation in the underfloor area is used for the information equipment room only. The ventilation system shall be so arranged, with approved smoke detection devices, that upon the detection of fire or products of combustion in the underfloor space the circulation of air will cease.
(4) Openings in raised floors for cables protect cables against abrasions and minimize the entrance of debris beneath the floor.
(5) Cables, other than those covered in (2) and those complying with (a), (b), and (c), shall be listed as Type DP cable having adequate fire-resistant characteristics suitable for use under raised floors of an information technology equipment room.

(a) Interconnecting cables enclosed in a raceway.

(b) Interconnecting cables listed with equipment manufactured prior to July 1, 1994, being installed with that equipment.

(c) Cable type designations Type TC (Article 336); Types CL2, CL3, and PLTC (Article 725); Type ITC (Article 727); Types NPLF and FPL (Article 760); Types OFC and OFN (Article 770); Types CM and MP (Article 800); and Type CATV (Article 820). These designations shall be permitted to have an additional letter P or R or G. Green, with one or more yellow stripes, insulated single conductor cables, 4 AWG and larger, marked "for use in cable trays" or "for CT use" shall be permitted for equipment grounding.

FPN: One method of defining fire resistance is by establishing that the cables do not spread fire to the top of the tray in the "Vertical Tray Flame Test" referenced in ANSI/UL 1581-1991, *Standard for Electrical Wires, Cables, and Flexible Cords.* Another method of defining fire resistance is for the damage (char length) not to exceed 1.5 m (4 ft 11 in.) when performing the CSA "Vertical Flame Test — Cables in Cable Trays," as described in CSA C22.2 No. 0.3-M-1985, *Test Methods for Electrical Wires and Cables.*

(6) Abandoned cables shall not be permitted to remain unless contained in metal raceways.

(E) Securing in Place. Power cables; communications cables; connecting cables; interconnecting cables; and associated boxes, connectors, plugs, and receptacles that are listed as part of, or for, information technology equipment shall not be required to be secured in place.

645.6 Cables Not in Information Technology Equipment Room. Cables extending beyond the information technology equipment room shall be subject to the applicable requirements of this *Code.*

FPN: For signaling circuits, refer to Article 725; for fiber optic circuits, refer to Article 770; and for communications circuits, refer to Article 800. For fire alarm systems, refer to Article 760.

645.7 Penetrations. Penetrations of the fire-resistant room boundary shall be in accordance with 300.21.

645.10 Disconnecting Means. A means shall be provided to disconnect power to all electronic equipment in the information technology equipment room. There shall also be a similar means to disconnect the power to all dedicated HVAC systems serving the room and cause all required fire/smoke dampers to close. The control for these disconnecting means shall be grouped and identified and shall be readily accessible at the principal exit doors. A single means to control both the electronic equipment and HVAC systems shall be permitted. Where a pushbutton is used as a means to disconnect power, pushing the button in shall disconnect the power.

Exception: Installations qualifying under the provisions of Article 685.

645.11 Uninterruptible Power Supplies (UPS). Unless otherwise permitted in (1) or (2), UPS systems installed within the information technology room, and their supply and output circuits, shall comply with 645.10. The disconnecting means shall also disconnect the battery from its load.

(1) Installations qualifying under the provisions of Article 685
(2) Power sources capable of supplying 750 volt-amperes or less derived either from UPS equipment or from battery circuits integral to electronic equipment

645.15 Grounding. All exposed non–current-carrying metal parts of an information technology system shall be grounded in accordance with Article 250 or shall be double insulated. Power systems derived within listed information technology equipment that supply information technology systems through receptacles or cable assemblies supplied as part of this equipment shall not be considered separately derived for the purpose of applying 250.20(D). Where signal reference structures are installed, they shall be bonded to the equipment grounding system provided for the information technology equipment.

> FPN No. 1: The bonding and grounding requirements in the product standards governing this listed equipment ensure that it complies with Article 250.

> FPN No. 2: Where isolated grounding-type receptacles are used, see 250.146(D) and 406.2(D).

645.16 Marking. Each unit of an information technology system supplied by a branch circuit shall be provided with a manufacturer's nameplate, which shall also include the input power requirements for voltage, frequency, and maximum rated load in amperes.

Class 1, Class 2, and Class 3 Remote-Control, Signaling, and Power-Limited Circuits

I. General

725.1 Scope. This article covers remote-control, signaling, and power-limited circuits that are not an integral part of a device or appliance.

> FPN: The circuits described herein are characterized by usage and electrical power limitations that differentiate them from electric light and power circuits; therefore, alternative requirements to those of Chapters 1 through 4 are given with regard to minimum wire sizes, derating factors, overcurrent protection, insulation requirements, and wiring methods and materials.

725.2 Definitions. For purposes of this article, the following definitions apply.

Abandoned Class 2, Class 3, and PLTC Cable. Installed Class 2, Class 3, and PLTC cable that is not terminated at equipment and not identified for future use with a tag.

Class 1 Circuit. The portion of the wiring system between the load side of the overcurrent device or power-limited supply and the connected equipment. The voltage and power limitations of the source are in accordance with 725.21.

Class 2 Circuit. The portion of the wiring system between the load side of a Class 2 power source and the connected equipment. Due to its power limitations, a Class 2 circuit considers safety from a fire initiation standpoint and provides acceptable protection from electric shock.

Class 3 Circuit. The portion of the wiring system between the load side of a Class 3 power source and the connected equipment. Due to its power limitations, a Class 3 circuit considers safety from a fire initiation standpoint. Since higher levels of voltage and current than Class 2 are permitted,

---

Source: NFPA 70, *National Electrical Code®*, NFPA, Quincy, MA, 2002 edition.

additional safeguards are specified to provide protection from an electric shock hazard that could be encountered.

725.3 Locations and Other Articles. Circuits and equipment shall comply with the articles or sections listed in 725.3(A) through (F). Only those sections of Article 300 referenced in this article shall apply to Class 1, Class 2, and Class 3 circuits.

(A) Number and Size of Conductors in Raceway. Section 300.17.

(B) Spread of Fire or Products of Combustion. Section 300.21. The accessible portion of abandoned Class 2, Class 3, and PLTC cables shall not be permitted to remain.

(C) Ducts, Plenums, and Other Air-Handling Spaces. Section 300.22 for Class 1, Class 2, and Class 3 circuits installed in ducts, plenums, or other space used for environmental air. Type CL2P or CL3P cables shall be permitted for Class 2 and Class 3 circuits.

(D) Hazardous (Classified) Locations. Articles 500 through 516 and Article 517, Part IV, where installed in hazardous (classified) locations.

(E) Cable Trays. Article 392, where installed in cable tray.

(F) Motor Control Circuits. Article 430, Part VI, where tapped from the load side of the motor branch-circuit protective device(s) as specified in 430.72(A).

725.5 Access to Electrical Equipment Behind Panels Designed to Allow Access. Access to electrical equipment shall not be denied by an accumulation of wires and cables that prevents removal of panels, including suspended ceiling panels.

725.6 Mechanical Execution of Work. Class 1, Class 2, and Class 3 circuits shall be installed in a neat and workmanlike manner. Cables and conductors installed exposed on the outer surface of ceiling and sidewalls shall be supported by structural components of the building in such a manner that the cable or conductors will not be damaged by normal building use. Such cables shall be attached to structural components by straps, staples, hangers, or similar fittings designed and installed so as not to damage the cable. The installation shall also conform with 300.4(D).

725.8 Safety-Control Equipment.

(A) Remote-Control Circuits. Remote-control circuits for safety-control equipment shall be classified as Class 1 if the failure of the equipment to operate introduces a direct fire or life hazard. Room thermostats, water temperature regulating devices, and similar controls used in conjunction with electrically controlled household heating and air conditioning shall not be considered safety-control equipment.

(B) Physical Protection. Where damage to remote-control circuits of safety control equipment would introduce a hazard, as covered in 725.8(A), all conductors of such remote-control circuits shall be installed in rigid metal conduit, intermediate metal conduit, rigid nonmetallic conduit, electrical metallic tubing, Type MI cable, Type MC cable, or be otherwise suitably protected from physical damage.

725.9 Class 1, Class 2, and Class 3 Circuit Grounding. Class 1, Class 2, and Class 3 circuits and equipment shall be grounded in accordance with Article 250.

725.10 Class 1, Class 2, and Class 3 Circuit Identification. Class 1, Class 2, and Class 3 circuits shall be identified at terminal and junction locations, in a manner that prevents unintentional interference with other circuits during testing and servicing.

725.15 Class 1, Class 2, and Class 3 Circuit Requirements. A remote-control, signaling, or power-limited circuit shall comply with the following parts of this article:

(1) Class 1 Circuits, Parts I and II
(2) Class 2 and Class 3 Circuits, Parts I and III

II. Class 1 Circuits

725.21 Class 1 Circuit Classifications and Power Source Requirements. Class 1 circuits shall be classified as either Class 1 power-limited circuits where they comply with the power limitations of 725.21(A) or as Class 1 remote-control and signaling circuits where they are used for remote control or signaling purposes and comply with the power limitations of 725.21(B).

(A) Class 1 Power-Limited Circuits. These circuits shall be supplied from a source that has a rated output of not more than 30 volts and 1000 volt-amperes.

(1) Class 1 Transformers. Transformers used to supply power-limited Class 1 circuits shall comply with Article 450.

(2) Other Class 1 Power Sources. Power sources other than transformers shall be protected by overcurrent devices rated at not more than 167 percent of the volt-ampere rating of the source divided by the rated voltage. The overcurrent devices shall not be interchangeable with overcurrent devices of higher ratings. The overcurrent device shall be permitted to be an integral part of the power supply.

To comply with the 1000 volt-ampere limitation of 725.21(A), the maximum output (VA_{max}) of power sources other than transformers shall be limited to 2500 volt-amperes, and the product of the maximum current (I_{max}) and maximum voltage (V_{max}) shall not exceed 10,000 volt-amperes. These ratings shall be determined with any overcurrent-protective device bypassed.

VA_{max} is the maximum volt-ampere output after one minute of operation regardless of load and with overcurrent protection bypassed, if used. Current-limiting impedance shall not be bypassed when determining VA_{max}.

I_{max} is the maximum output current under any noncapacitive load, including short circuit, and with overcurrent protection bypassed, if used. Current-limiting impedance should not be bypassed when determining I_{max}. Where a current-limiting impedance, listed for the purpose or as part of a listed product, is used in combination with a stored energy source, for example, storage battery, to limit the output current, I_{max} limits apply after 5 seconds.

V_{max} is the maximum output voltage regardless of load with rated input applied.

(B) Class 1 Remote-Control and Signaling Circuits. These circuits shall not exceed 600 volts. The power output of the source shall not be required to be limited.

725.23 Class 1 Circuit Overcurrent Protection. Overcurrent protection for conductors 14 AWG and larger shall be provided in accordance with the conductor ampacity, without applying the derating factors of 310.15 to the ampacity calculation. Overcurrent protection shall not exceed 7 amperes for 18 AWG conductors and 10 amperes for 16 AWG.

Exception: Where other articles of this Code permit or require other overcurrent protection.

FPN: For example, see 430.72 for motors, 610.53 for cranes and hoists, and 517.74(B) and 660.9 for X-ray equipment.

725.24 Class 1 Circuit Overcurrent Device Location. Overcurrent devices shall be located as specified in 725.24(A) through (E).

(A) Point of Supply. Overcurrent devices shall be located at the point where the conductor to be protected receives its supply.

(B) Feeder Taps. Class 1 circuit conductors shall be permitted to be tapped, without overcurrent protection at the tap, where the overcurrent device protecting the circuit conductor is sized to protect the tap conductor.

(C) Transformer Taps. Class 1 circuit conductors 14 AWG and larger that are tapped from the load side of the overcurrent-protective device(s) of a controlled light and power circuit shall require only short-circuit and ground-fault protection and shall be permitted to be protected by the branch-circuit overcurrent protective device(s) where the rating of the protective device(s) is not more than 300 percent of the ampacity of the Class 1 circuit conductor.

(D) Primary Side of Transformer. Class 1 circuit conductors supplied by the secondary of a single-phase transformer having only a 2-wire (single-voltage) secondary shall be permitted to be protected by overcurrent protection provided on the primary side of the transformer, provided this protection is in accordance with 450.3 and does not exceed the value determined by multiplying the secondary conductor ampacity by the secondary-to-primary transformer voltage ratio. Transformer secondary conductors other than 2 wire shall not be considered to be protected by the primary overcurrent protection.

(E) Input Side of Electronic Power Source. Class 1 circuit conductors supplied by the output of a single-phase, listed electronic power source, other than a transformer, having only a 2-wire (single voltage) output for connection to Class 1 circuits shall be permitted to be protected by overcurrent protection provided on the input side of the electronic power source, provided this protection does not exceed the value determined by multiplying the Class 1 circuit conductor ampacity by the output-to-input voltage ratio. Electronic power source outputs, other than 2 wire (single voltage), shall not be considered to be protected by the primary overcurrent protection.

725.25 Class 1 Circuit Wiring Methods. Installations of Class 1 circuits shall be in accordance with Article 300 and the other appropriate articles in Chapter 3.

Exception No. 1: The provisions of 725.26 through 725.28 shall be permitted to apply in installations of Class 1 circuits.

Exception No. 2: Methods permitted or required by other articles of this Code shall apply to installations of Class 1 circuits.

725.26 Conductors of Different Circuits in the Same Cable, Cable Tray, Enclosure, or Raceway. Class 1 circuits shall be permitted to be installed with other circuits as specified in 725.26(A) and (B).

(A) Two or More Class 1 Circuits. Class 1 circuits shall be permitted to occupy the same cable, cable tray, enclosure, or raceway without regard to whether the individual circuits are alternating current or direct current, provided all conductors are insulated for the maximum voltage of any conductor in the cable, cable tray, enclosure, or raceway.

(B) Class 1 Circuits with Power Supply Circuits. Class 1 circuits shall be permitted to be installed with power supply conductors as specified in 725.26(B)(1) through (B)(4).

(1) In a Cable, Enclosure, or Raceway. Class 1 circuits and power supply circuits shall be permitted to occupy the same cable, enclosure, or raceway only where the equipment powered is functionally associated.

(2) In Factory- or Field-Assembled Control Centers. Class 1 circuits and power supply circuits shall be permitted to be installed in factory- or field-assembled control centers.

(3) In a Manhole. Class 1 circuits and power supply circuits shall be permitted to be installed as underground conductors in a manhole in accordance with one of the following:

(1) The power-supply or Class 1 circuit conductors are in a metal-enclosed cable or Type UF cable.
(2) The conductors are permanently separated from the power-supply conductors by a continuous firmly fixed nonconductor, such as flexible tubing, in addition to the insulation on the wire.
(3) The conductors are permanently and effectively separated from the power supply conductors and securely fastened to racks, insulators, or other approved supports.

(4) In cable trays, where the Class 1 circuit conductors and power-supply conductors not functionally associated with them are separated by a solid fixed barrier of a material compatible with the cable tray, or where the power-supply or Class 1 circuit conductors are in a metal-enclosed cable.

725.27 Class 1 Circuit Conductors.

(A) Sizes and Use. Conductors of sizes 18 AWG and 16 AWG shall be permitted to be used, provided they supply loads that do not exceed the ampacities given in 402.5 and are installed in a raceway, an approved enclosure, or a listed cable. Conductors larger than 16 AWG shall not supply loads greater than the ampacities given in 310.15. Flexible cords shall comply with Article 400.

(B) Insulation. Insulation on conductors shall be suitable for 600 volts. Conductors larger than 16 AWG shall comply with Article 310. Conductors in sizes 18 AWG and 16 AWG shall be Type FFH-2, KF-2, KFF-2, PAF, PAFF, PF, PFF, PGF, PGFF, PTF, PTFF, RFH-2, RFHH-2, RFHH-3, SF-2, SFF-2, TF,

TFF, TFFN, TFN, ZF, or ZFF. Conductors with other types and thicknesses of insulation shall be permitted if listed for Class 1 circuit use.

725.28 Number of Conductors in Cable Trays and Raceway, and Derating.

(A) Class 1 Circuit Conductors. Where only Class 1 circuit conductors are in a raceway, the number of conductors shall be determined in accordance with 300.17. The derating factors given in 310.15(B)(2)(a) shall apply only if such conductors carry continuous loads in excess of 10 percent of the ampacity of each conductor.

(B) Power-Supply Conductors and Class 1 Circuit Conductors. Where power-supply conductors and Class 1 circuit conductors are permitted in a raceway in accordance with 725.26, the number of conductors shall be determined in accordance with 300.17. The derating factors given in 310.15(B)(2)(a) shall apply as follows:

(1) To all conductors where the Class 1 circuit conductors carry continuous loads in excess of 10 percent of the ampacity of each conductor and where the total number of conductors is more than three
(2) To the power-supply conductors only, where the Class 1 circuit conductors do not carry continuous loads in excess of 10 percent of the ampacity of each conductor and where the number of power-supply conductors is more than three

(C) Class 1 Circuit Conductors in Cable Trays. Where Class 1 circuit conductors are installed in cable trays, they shall comply with the provisions of 392.9 through 392.11.

725.29 Circuits Extending Beyond One Building. Class 1 circuits that extend aerially beyond one building shall also meet the requirements of Article 225.

III. Class 2 and Class 3 Circuits

725.41 Power Sources for Class 2 and Class 3 Circuits.

(A) Power Source. The power source for a Class 2 or a Class 3 circuit shall be as specified in 725.41(A)(1), (2), (3), (4), or (5):

FPN No. 1: Figure 725.41 illustrates the relationships between Class 2 or Class 3 power sources, their supply, and the Class 2 or Class 3 circuits.

FPN No. 2: Table 11(A) and Table 11(B) in Chapter 9 provide the requirements for listed Class 2 and Class 3 power sources.

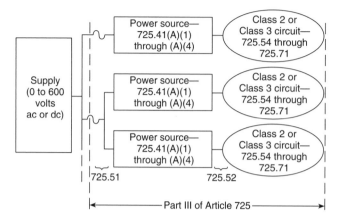

Figure 725.41 Class 2 and Class 3 circuits.

(1) A listed Class 2 or Class 3 transformer
(2) A listed Class 2 or Class 3 power supply
(3) Other listed equipment marked to identify the Class 2 or Class 3 power source

Exception: Thermocouples shall not require listing as a Class 2 power source.

FPN: Examples of other listed equipment are as follows:

(1) A circuit card listed for use as a Class 2 or Class 3 power source where used as part of a listed assembly
(2) A current-limiting impedance, listed for the purpose, or part of a listed product, used in conjunction with a non–power-limited transformer or a stored energy source, for example, storage battery, to limit the output current
(3) A thermocouple

(4) Listed information technology (computer) equipment limited power circuits.

FPN: One way to determine applicable requirements for listing of information technology (computer) equipment is to refer to UL 1950-1995, *Standard for Safety of Information Technology Equipment, Including Electrical Business Equipment.* Typically such circuits are used to interconnect information technology equipment for the purpose of exchanging information (data).

(5) A dry cell battery shall be considered an inherently limited Class 2 power source, provided the voltage is 30 volts or less and the capacity is equal to or less than that available from series connected No. 6 carbon zinc cells.

(B) Interconnection of Power Sources. Class 2 or Class 3 power sources shall not have the output connections paralleled or otherwise interconnected unless listed for such interconnection.

725.42 Circuit Marking. The equipment shall be durably marked where plainly visible to indicate each circuit that is a Class 2 or Class 3 circuit.

725.51 Wiring Methods on Supply Side of the Class 2 or Class 3 Power Source. Conductors and equipment on the supply side of the power source shall be installed in accordance with the appropriate requirements of Chapters 1 through 4. Transformers or other devices supplied from electric light or power circuits shall be protected by an overcurrent device rated not over 20 amperes.

Exception: The input leads of a transformer or other power source supplying Class 2 and Class 3 circuits shall be permitted to be smaller than 14 AWG, but not smaller than 18 AWG if they are not over 12 in. (305 mm) long and if they have insulation that complies with 725.27(B).

725.52 Wiring Methods and Materials on Load Side of the Class 2 or Class 3 Power Source. Class 2 and Class 3 circuits on the load side of the power source shall be permitted to be installed using wiring methods and materials in accordance with either 725.52(A) or (B).

(A) Class 1 Wiring Methods and Materials. Installation shall be in accordance with 725.25.

Exception No. 1: The derating factors that are given in 310.15(B)(2)(a) shall not apply.

Exception No. 2: Class 2 and Class 3 circuits shall be permitted to be reclassified and installed as Class 1 circuits if the Class 2 and Class 3 markings required in 725.42 are eliminated and the entire circuit is installed using the wiring methods and materials in accordance with Part II, Class 1 circuits.

> FPN: Class 2 and Class 3 circuits reclassified and installed as Class 1 circuits are no longer Class 2 or Class 3 circuits, regardless of the continued connection to a Class 2 or Class 3 power source.

(B) Class 2 and Class 3 Wiring Methods. Conductors on the load side of the power source shall be insulated at not less than the requirements of 725.71 and shall be installed in accordance with 725.54 and 725.61.

Exception No. 1: As provided for in 620.21 for elevators and similar equipment.

Exception No. 2: Other wiring methods and materials installed in accordance with the requirements of 725.3 shall be permitted to extend or replace the conductors and cables described in 725.71 and permitted by 725.52(B).

725.54 Installation of Conductors and Equipment in Cables, Compartments, Cable Trays, Enclosures, Manholes, Outlet Boxes, Device Boxes, and Raceways for Class 2 and Class 3 Circuits. Conductors and equipment for Class 2 and Class 3 circuits shall be installed in accordance with 725.55 through 725.58.

725.55 Separation from Electric Light, Power, Class 1, Non–Power-Limited Fire Alarm Circuit Conductors, and Medium Power Network-Powered Broadband Communications Cables.

(A) General. Cables and conductors of Class 2 and Class 3 circuits shall not be placed in any cable, cable tray, compartment, enclosure, manhole, outlet box, device box, raceway, or similar fitting with conductors of electric light, power, Class 1, non–power-limited fire alarm circuits, and medium power network-powered broadband communications circuits unless permitted by 725.55(B) through (J).

(B) Separated by Barriers. Class 2 and Class 3 circuits shall be permitted to be installed together with Class 1, non–power-limited fire alarm and medium power network-powered broadband communications circuits where they are separated by a barrier.

(C) Raceways Within Enclosures. In enclosures, Class 2 and Class 3 circuits shall be permitted to be installed in a raceway to separate them from Class 1, non–power-limited fire alarm and medium power network-powered broadband communications circuits.

(D) Associated Systems Within Enclosures. Class 2 and Class 3 circuit conductors in compartments, enclosures, device boxes, outlet boxes, or similar fittings shall be permitted to be installed with electric light, power, Class 1, non–power-limited fire alarm, and medium power network-powered broadband communications circuits where they are introduced solely to connect the equipment connected to Class 2 and Class 3 circuits, and where (1) or (2) applies:

(1) The electric light, power, Class 1, non–power-limited fire alarm, and medium power network-powered broadband communications circuit conductors are routed to maintain a minimum of 6 mm (0.25 in.) separation from the conductors and cables of Class 2 and Class 3 circuits.
(2) The circuit conductors operate at 150 volts or less to ground and also comply with one of the following:

    a. The Class 2 and Class 3 circuits are installed using Type CL3, CL3R, or CL3P or permitted substitute cables, provided these Class 3 cable conductors extending beyond the jacket are separated by a minimum of 6 mm (0.25 in.) or by a nonconductive sleeve or nonconductive barrier from all other conductors.

b. The Class 2 and Class 3 circuit conductors are installed as a Class 1 circuit in accordance with 725.21.

(E) Enclosures with Single Opening. Class 2 and Class 3 circuit conductors entering compartments, enclosures, device boxes, outlet boxes, or similar fittings shall be permitted to be installed with Class 1, non–power-limited fire alarm and medium power network-powered broadband communications circuits where they are introduced solely to connect the equipment connected to Class 2 and Class 3 circuits. Where Class 2 and Class 3 circuit conductors must enter an enclosure that is provided with a single opening, they shall be permitted to enter through a single fitting (such as a tee), provided the conductors are separated from the conductors of the other circuits by a continuous and firmly fixed nonconductor, such as flexible tubing.

(F) Manholes. Underground Class 2 and Class 3 circuit conductors in a manhole shall be permitted to be installed with Class 1, non–power-limited fire alarm and medium power network-powered broadband communications circuits where one of the following conditions is met:

(1) The electric light, power, Class 1, non–power-limited fire alarm and medium power network-powered broadband communications circuit conductors are in a metal-enclosed cable or Type UF cable.
(2) The Class 2 and Class 3 circuit conductors are permanently and effectively separated from the conductors of other circuits by a continuous and firmly fixed nonconductor, such as flexible tubing, in addition to the insulation or covering on the wire.
(3) The Class 2 and Class 3 circuit conductors are permanently and effectively separated from conductors of the other circuits and securely fastened to racks, insulators, or other approved supports.

(G) Article 780. Class 2 and Class 3 conductors as permitted by 780.6(A) shall be permitted to be installed in accordance with Article 780.

(H) Cable Trays. Class 2 and Class 3 circuit conductors shall be permitted to be installed in cable trays, where the conductors of the electric light, Class 1, and non–power-limited fire alarm circuits are separated by a solid fixed barrier of a material compatible with the cable tray or where the Class 2 or Class 3 circuits are installed in Type MC cable.

(I) In Hoistways. In hoistways, Class 2 or Class 3 circuit conductors shall be installed in rigid metal conduit, rigid nonmetallic conduit, intermediate metal conduit, liquidtight flexible nonmetallic conduit, or electrical metallic tubing. For elevators or similar equipment, these conductors shall be permitted to be installed as provided in 620.21.

(J) Other Applications. For other applications, conductors of Class 2 and Class 3 circuits shall be separated by at least 50 mm (2 in.) from conductors of any electric light, power, Class 1 non–power-limited fire alarm or medium power network-powered broadband communications circuits unless one of the following conditions is met:

(1) Either (a) all of the electric light, power, Class 1, non–power-limited fire alarm and medium power network-powered broadband communications circuit conductors or (b) all of the Class 2 and Class 3 circuit conductors are in a raceway or in metal-sheathed, metal-clad, non–metallic-sheathed, or Type UF cables.

(2) All of the electric light, power, Class 1 non–power-limited fire alarm, and medium power network-powered broadband communications circuit conductors are permanently separated from all of the Class 2 and Class 3 circuit conductors by a continuous and firmly fixed non-conductor, such as porcelain tubes or flexible tubing, in addition to the insulation on the conductors.

725.56 Installation of Conductors of Different Circuits in the Same Cable, Enclosure, or Raceway.

(A) Two or More Class 2 Circuits. Conductors of two or more Class 2 circuits shall be permitted within the same cable, enclosure, or raceway.

(B) Two or More Class 3 Circuits. Conductors of two or more Class 3 circuits shall be permitted within the same cable, enclosure, or raceway.

(C) Class 2 Circuits with Class 3 Circuits. Conductors of one or more Class 2 circuits shall be permitted within the same cable, enclosure, or raceway with conductors of Class 3 circuits, provided that the insulation of the Class 2 circuit conductors in the cable, enclosure, or raceway is at least that required for Class 3 circuits.

(D) Class 2 and Class 3 Circuits with Communications Circuits.

(1) Classified as Communications Circuits. Class 2 and Class 3 circuit conductors shall be permitted in the same cable with communications circuits, in which case the Class 2 and Class 3 circuits shall be classified as communications circuits and shall be installed in accordance with the requirements of Article 800. The cables shall be listed as communications cables or multipurpose cables.

(2) Composite Cables. Cables constructed of individually listed Class 2, Class 3, and communications cables under a common jacket shall be permitted to be classified as communications cables. The fire resistance rating of the composite cable shall be determined by the performance of the composite cable.

(E) Class 2 or Class 3 Cables with Other Circuit Cables. Jacketed cables of Class 2 or Class 3 circuits shall be permitted in the same enclosure or raceway with jacketed cables of any of the following:

(1) Power-limited fire alarm systems in compliance with Article 760
(2) Nonconductive and conductive optical fiber cables in compliance with Article 770
(3) Communications circuits in compliance with Article 800
(4) Community antenna television and radio distribution systems in compliance with Article 820
(5) Low-power, network-powered broadband communications in compliance with Article 830

725.57 Installation of Circuit Conductors Extending Beyond One Building. Where Class 2 or Class 3 circuit conductors extend beyond one building and are run so as to be subject to accidental contact with electric light or power conductors operating over 300 volts to ground, or are exposed to lightning on interbuilding circuits on the same premises, the requirements of the following shall also apply:

(1) Sections 800.10, 800.12, 800.13, 800.31, 800.32, 800.33, and 800.40 for other than coaxial conductors
(2) Sections 820.10, 820.33, and 820.40 for coaxial conductors

725.58 Support of Conductors. Class 2 or Class 3 circuit conductors shall not be strapped, taped, or attached by any means to the exterior of any conduit or other raceway as a means of support. These conductors shall be permitted to be installed as permitted by 300.11(B)(2).

725.61 Applications of Listed Class 2, Class 3, and PLTC Cables. Class 2, Class 3, and PLTC cables shall comply with any of the requirements described in 725.61(A) through (F).

(A) Plenum. Cables installed in ducts, plenums, and other spaces used for environmental air shall be Type CL2P or CL3P. Abandoned cables shall not be permitted to remain. Listed wires and cables installed in compliance with 300.22 shall be permitted.

(B) Riser. Cables installed in risers shall be as described in any of (1), (2), or (3):

(1) Cables installed in vertical runs and penetrating more than one floor, or cables installed in vertical runs in a shaft, shall be Type CL2R or CL3R. Floor penetrations requiring Type CL2R or CL3R shall contain only cables suitable for riser or plenum use. Abandoned cables shall not be permitted to remain.
(2) Other cables as covered in Table 725.61 and other listed wiring methods as covered in Chapter 3 shall be installed in metal raceways or located in a fireproof shaft having firestops at each floor.
(3) Type CL2, CL3, CL2X, and CL3X cables shall be permitted in one- and two-family dwellings.

FPN: See 300.21 for firestop requirements for floor penetrations.

(C) Cable Trays. Cables installed in cable trays outdoors shall be Type PLTC. Cables installed in cable trays indoors shall be Types PLTC, CL3P, CL3R, CL3, CL2P, CL2R, and CL2.

FPN: See 800.52(D) for cables permitted in cable trays.

(D) Hazardous (Classified) Locations. Cables installed in hazardous locations shall be as described in 725.61(D)(1) through (D)(4).

(1) Type PLTC. Cables installed in hazardous (classified) locations shall be Type PLTC. Where the use of Type PLTC cable is permitted by 501.4(B), 502.4(B), and 504.20, the cable shall be installed in cable trays, in raceways supported by messenger wire, or otherwise adequately supported and mechanically protected by angles, struts, channels, or other mechanical means. The cable shall be permitted to be directly buried where the cable is listed for this use.

(2) Nonincendive Field Wiring. Wiring for Class 2 circuits as permitted by 501.4(B)(3) shall be permitted.

(3) Thermocouple Circuits. Conductors in Type PLTC cables used for Class 2 thermocouple circuits shall be permitted to be any of the materials used for thermocouple extension wire.

(4) In Industrial Establishments. In industrial establishments where the conditions of maintenance and supervision ensure that only qualified persons service the installation, and where the cable is not subject to physical damage, Type PLTC cable that complies with the crush and impact requirements of Type MC cable and is identified for such use shall be permitted as open wiring between cable tray and utilization equipment in lengths not to exceed 15 m (50 ft). The cable shall be supported and protected against physical damage using mechanical protection such as dedicated struts, angles, or channels. The cable shall be supported and secured at intervals not exceeding 1.75 m (6 ft).

(E) Other Wiring Within Buildings. Cables installed in building locations other than those covered in 725.61(A) through (D) shall be as described in any of (1) through (6). Abandoned cables in hollow spaces shall not be permitted to remain.

(1) Type CL2 or CL3 shall be permitted.
(2) Type CL2X or CL3X shall be permitted to be installed in a raceway or in accordance with other wiring methods covered in Chapter 3.
(3) Cables shall be permitted to be installed in nonconcealed spaces where the exposed length of cable does not exceed 3 m (10 ft).

(4) Listed Type CL2X cables less than 6 mm (0.25 in.) in diameter and listed Type CL3X cables less than 6 mm (0.25 in.) in diameter shall be permitted to be installed in one- and two-family dwellings.
(5) Listed Type CL2X cables less than 6 mm (0.25 in.) in diameter and listed Type CL3X cables less than 6 mm (0.25 in.) in diameter shall be permitted to be installed in nonconcealed spaces in multifamily dwellings.
(6) Type CMUC undercarpet communications wires and cables shall be permitted to be installed under carpet.

(F) Cross-Connect Arrays. Type CL2 or CL3 conductors or cables shall be used for cross-connect arrays.

(G) Class 2 and Class 3 Cable Uses and Permitted Substitutions. The uses and permitted substitutions for Class 2 and Class 3 cables listed in Table 725.61 shall be considered suitable for the purpose and shall be permitted.

FPN: For information on Types CMP, CMR, CH, and CMX cables, see 800.51.

Table 725.61   Cable Uses and Permitted Substitutions

Cable Type	Use	References	Permitted Substitutions
CL3P	Class 3 plenum cable	725.61(A)	CMP
CL2P	Class 2 plenum cable	725.61(A)	CMP, CL3P
CL3R	Class 3 riser cable	725.61(B)	CMP, CL3P, CMR
CL2R	Class 2 riser cable	725.61(B)	CMP, CL3P, CL2P, CMR, CL3R
PLTC	Power-limited tray cable	725.61(C) and (D)	
CL3	Class 3 cable	725.61(B), (E), and (F)	CMP, CL3P, CMR, CL3R, CMG, CM, PLTC
CL2	Class 2 cable	725.61(B), (E), and (F)	CMP, CL3P, CL2P, CMR, CL3R, CL2R, CMG, CM, PLTC, CL3
CL3X	Class 3 cable, limited use	725.61(B) and (E)	CMP, CL3P, CMR, CL3R, CMG, CM, PLTC, CL3, CMX
CL2X	Class 2 cable, limited use	725.61(B) and (E)	CMP, CL3P, CL2P, CMR, CL3R, CL2R, CMG, CM, PLTC, CL3, CL2, CMX, CL3X

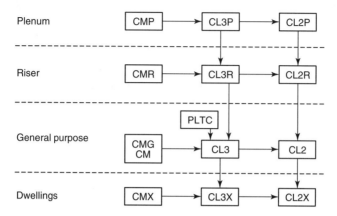

Type CM—Communications wires and cables
Type CL2 and CL3—Class 2 and Class 3 remote-control, signaling, and power-limited cables
Type PLTC—Power-limited tray cable

A ▸ B   Cable A shall be permitted to be used in place of cable B.

Figure 725.61 Cable substitution hierarchy

725.71 Listing and Marking of Class 2, Class 3, and Type PLTC Cables. Class 2, Class 3, and Type PLTC cables installed as wiring within buildings shall be listed as being resistant to the spread of fire and other criteria in accordance with 725.71(A) through (G) and shall be marked in accordance with 725.71(H).

(A) Types CL2P and CL3P. Types CL2P and CL3P plenum cables shall be listed as being suitable for use in ducts, plenums, and other space used for environmental air and shall also be listed as having adequate fire-resistant and low smoke-producing characteristics.

> FPN: One method of defining *low smoke-producing cable* is by establishing an acceptable value of the smoke produced when tested in accordance with NFPA 262-1999, *Standard Method of Test for Flame Travel and Smoke of Wires and Cables for Use in Air-Handling Spaces*, to a maximum peak optical density of 0.5 and a maximum average optical density of 0.15. Similarly, one method of defining fire-resistant cables is by establishing a maximum allowable flame travel distance of 1.52 m (5 ft) when tested in accordance with the same test.

(B) Types CL2R and CL3R. Types CL2R and CL3R riser cables shall be listed as being suitable for use in a vertical run in a shaft or from floor to floor and shall also be listed as having fire-resistant characteristics capable of preventing the carrying of fire from floor to floor.

FPN: One method of defining fire-resistant characteristics capable of preventing the carrying of fire from floor to floor is that the cables pass the requirements of ANSI/UL 1666-1997, *Test for Flame Propagation Height of Electrical and Optical-Fiber Cable Installed Vertically in Shafts.*

(C) Types CL2 and CL3. Types CL2 and CL3 cables shall be listed as being suitable for general-purpose use, with the exception of risers, ducts, plenums, and other space used for environmental air and shall also be listed as being resistant to the spread of fire.

FPN: One method of defining *resistant to the spread of fire* is that the cables do not spread fire to the top of the tray in the vertical tray flame test in ANSI/UL 1581-1991, *Reference Standard for Electrical Wires, Cables and Flexible Cords.*

Another method of defining *resistant to the spread of fire* is for the damage (char length) not to exceed 1.5 m (4 ft 11 in.) when performing the CSA vertical flame test for cables in cable trays, as described in CSA C22.2 No. 0.3-M-1985, *Test Methods for Electrical Wires and Cables.*

(D) Types CL2X and CL3X. Types CL2X and CL3X limited-use cables shall be listed as being suitable for use in dwellings and for use in raceway and shall also be listed as being resistant to flame spread.

FPN: One method of determining that cable is resistant to flame spread is by testing the cable to the VW-1 (vertical-wire) flame test in ANSI/UL 1581-1991, *Reference Standard for Electrical Wires, Cables and Flexible Cords.*

(E) Type PLTC. Type PLTC nonmetallic-sheathed, power-limited tray cable shall be listed as being suitable for cable trays and shall consist of a factory assembly of two or more insulated conductors under a nonmetallic jacket. The insulated conductors shall be 22 AWG through 12 AWG. The conductor material shall be copper (solid or stranded). Insulation on conductors shall be suitable for 300 volts. The cable core shall be either (1) two or more parallel conductors, (2) one or more group assemblies of twisted or parallel conductors, or (3) a combination thereof. A metallic shield or a metallized foil shield with drain wire(s) shall be permitted to be applied either over the cable core, over groups of conductors, or both. The cable shall be listed as being resistant to the spread of fire. The outer jacket shall be a sunlight- and moisture-resistant nonmetallic material.

Exception No. 1: Where a smooth metallic sheath, continuous corrugated metallic sheath, or interlocking tape armor is applied over the nonmetallic jacket, an overall nonmetallic jacket shall not be required. On metallic-sheathed cable without an overall nonmetallic jacket, the information required in 310.11 shall be located on the nonmetallic jacket under the sheath.

Exception No. 2: Conductors in PLTC cables used for Class 2 thermocouple circuits shall be permitted to be any of the materials used for thermocouple extension wire.

> FPN: One method of defining *resistant to the spread of fire* is that the cables do not spread fire to the top of the tray in the vertical tray flame test in ANSI/UL 1581-1991, *Reference Standard for Electrical Wires, Cables and Flexible Cords.*
>
> Another method of defining *resistant to the spread of fire* is for the damage (char length) not to exceed 1.5 m (4 ft 11 in.) when performing the CSA vertical flame test for cables in cable trays, as described in CSA C22.2 No. 0.3-M-1985, *Test Methods for Electrical Wires and Cables.*

(F) Class 2 and Class 3 Cable Voltage Ratings. Class 2 cables shall have a voltage rating of not less than 150 volts. Class 3 cables shall have a voltage rating of not less than 300 volts.

(G) Class 3 Single Conductors. Class 3 single conductors used as other wiring within buildings shall not be smaller than 18 AWG and shall be Type CL3. Conductor types described in 725.27(B) that are also listed as Type CL3 shall be permitted.

> FPN: One method of defining *resistant to the spread of fire* is that the cables do not spread fire to the top of the tray in the vertical tray flame test in ANSI/UL 1581-1991, *Reference Standard for Electrical Wires, Cables and Flexible Cords.*
>
> Another method of defining *resistant to the spread of fire* is for the damage (char length) not to exceed 1.5 m (4 ft 11 in.) when performing the CSA vertical flame test for cables in cable trays as described in CSA C22.2 No. 0.3-M-1985, *Test Methods for Electrical Wires and Cables.*

(H) Marking. Cables shall be marked in accordance with 310.11(A)(2), (3), (4), and (5) and Table 725.71. Voltage ratings shall not be marked on the cables.

> FPN: Voltage markings on cables may be misinterpreted to suggest that the cables may be suitable for Class 1 electric light and power applications.

Exception: Voltage markings shall be permitted where the cable has multiple listings and a voltage marking is required for one or more of the listings.

> FPN: Class 2 and Class 3 cable types are listed in descending order of fire resistance rating, and Class 3 cables are listed above Class 2 cables, because Class 3 cables can substitute for Class 2 cables.

Table 725.71   Cable Markings

Cable Marking	Type	Listing References
CL3P	Class 3 plenum cable	725.71(A), (F), and (H)
CL2P	Class 2 plenum cable	725.71(A) and (H)
CL3R	Class 3 riser cable	725.71(B), (F), and (H)
CL2R	Class 2 riser cable	725.71(B) and (H)
PLTC	Power-limited tray cable	725.71(E) and (H)
CL3	Class 3 cable	725.71(C), (F), and (H)
CL2	Class 2 cable	725.71(C), (F), and (H)
CL3X	Class 3 cable, limited use	725.71(D), (F), and (H)
CL2X	Class 2 cable, limited use	725.71(D), (F), and (H)

CHAPTER

Fire Alarm Circuits and Systems

Like most articles in the *National Electrical Code®* (*NEC®*), Article 760 begins with a statement about the scope of the article. Section 760.1 says that Article 760 applies to the "installation of wiring and equipment of fire alarm systems including all circuits controlled and powered by the fire alarm system." Then follows a Fine Print Note (FPN) describing what types of systems and circuits are included under "fire alarm systems" and "circuits controlled and powered by the fire alarm system." The FPN states that fire alarm systems include fire detection, alarm notification, guard's tour, sprinkler waterflow, and sprinkler supervisory systems. Circuits controlled and powered by the fire alarm system may perform many functions, but the circuits are covered by Article 760 only if they are controlled and powered from the fire alarm system. Some of the fire alarm circuit functions may include elevator capture and shutdown, the control of doors and smoke dampers, and control of fans. (The FPN only mentions fan shutdown, but the list is not all-inclusive, and some fans, such as smoke exhaust fans and fans for stairwell pressurization, will be started when a fire is detected.) Many of the functions are dictated by building codes, and some of the functions may be required in buildings where there is no "fire alarm system."

For example, in some buildings, other codes require the capture of elevators, but do not require full fire alarm systems and may not require notification of all the building occupants, so individual residential-type smoke detectors are sometimes used. Depending on local codes and local interpretations, elevator circuits may be covered only under the general requirements of the *Code* or under either Article 725 or Article 760.

The FPN to Section 760.1 also directs the reader to *NFPA 72®*, the *National Fire Alarm Code®*, for additional information on the "monitoring for integrity requirements of fire alarm systems." This information helps to put Article 760 in proper context. As previously noted, Article 760 covers installation of equipment and wiring for fire alarm systems. The *NEC* and Article 760 do not say where a fire alarm system should be installed or how it should be designed. Nor, as implied by the reference to *NFPA 72*, does the *NEC* cover the performance requirements of a fire alarm system.

Electricians doing residential wiring often ask where it says in the *NEC* that they have to install smoke detectors in dwellings. The *NEC* does not cover this subject. The requirements for smoke detectors and fire alarm systems depend on what codes are adopted and enforced in any particular location, but are typically covered by NFPA *101®*, which is the *Life Safety Code®*, or other building codes. In the case of residential smoke detectors, the building codes may say precisely where the detectors are required and how the occupants are to be notified. In other occupancies, the building codes may only say which occupancies must have fire alarm systems and may refer to the *National Fire Alarm Code* for the design and operational requirements for the systems. "Household fire warning systems" (which include a control panel) and the "single or multiple station smoke alarms" commonly used in dwelling units are also covered by *NFPA 72*. Once the system and wiring have been designed and the equipment and devices have been selected, the *NEC* is used for the installation of the system.

NFPA 72 (1999, p. 15) defines a fire alarm system as

A system or portion of a combination system that consists of components and circuits arranged to monitor and annunciate the status of fire alarm or supervisory signal-initiating devices and to initiate the appropriate response to those signals.

Note that based on this definition, a fire alarm system need not have any smoke detectors, horns, or strobes if the system is used only to supervise a sprinkler system. However, in order to be listed and considered as a fire alarm system, the components of the system must be connected to and monitored by a central panel such as the one shown in Figure 4.1. Article 760 is primarily concerned with those systems that have a central control panel and not with interconnected stand-alone detectors such as those commonly used in new single-family homes. If specifically listed fire alarm circuit cables are to be used to interconnect such residential detectors, the wiring should be selected according to the detector manufacturer's instructions.

Although Section 760.1 says that Article 760 covers the installation of wiring and equipment, most of Article 760 is concerned with what many electricians and installers would consider wiring and not with the installation of control panels and devices. Control panels

FIGURE 4.1 Fire alarm control panels.

and devices are covered quite well by Section 110.3(B), which requires that listed equipment be installed according to the instructions. However, the term *equipment* as used in the *NEC* is defined in Article 100 and includes all types of "material, fittings, devices, appliances, luminaires (fixtures), apparatus, and the like used as part of, or in connection with, an electrical installation." So equipment includes all the wiring methods that may be used in an installation, and these wiring methods are the primary focus of Article 760.

Until the 1987 *NEC*, Article 760 contained a requirement that fire alarm circuits be "supervised" to ensure that the system would function when needed and to alert a user if a problem developed in a circuit. But this rule was not an installation requirement. Circuit supervision, now called "monitoring for integrity," is covered by *NFPA 72* because the requirement for monitoring for integrity is not primarily an installation issue; it is related more to the design, selection, and performance of circuits and equipment. Thus, the reference in Section 760.1 FPN to *NFPA 72* for "monitoring for integrity requirements." Monitoring for integrity of fire alarm circuits is covered in more detail under the heading "Related Requirements" near the end of this chapter.

The installation of fire alarm circuits is the primary focus of this chapter. Fire alarm circuits are defined in Section 760.2 as

> The portion of the wiring system between the load side of the overcurrent device or the power-limited supply and the connected equipment for all circuits powered by the fire alarm system. Fire alarm circuits are classified as either non-power-limited or power-limited.

Although this book is mostly concerned with power-limited circuits, this chapter also covers non-power-limited circuits for fire alarm systems.

CIRCUIT CLASSIFICATIONS

According to the definition of fire alarm circuits given in Section 760.2, the circuits are classified either as non-power-limited or power-limited. Non-power-limited fire alarm circuits are abbreviated in Article 760 by the acronym NPLFA; it follows that the acronym for power-limited fire alarm circuits is PLFA. These shortcuts are used extensively in Article 760, even though they are often spelled out in other locations in the *NEC*, and the acronyms are also used in this book.

The following discussion covers the differences between PLFA and NPLFA circuits. However, many fire alarm control panels use both types of circuits and power supplies. Often the detection devices and their "initiating device circuits" are PLFA circuits, whereas the horns and strobes are supplied by NPLFA "notification appliance circuits." Both types of power supplies will be included in such control panels, and the panels or terminals will be marked to indicate the classes of circuits used.

A *device* is defined in Article 100 as "A unit of an electrical system that is intended to carry but not utilize electrical energy." An *appliance* is defined as "Utilization equipment . . . installed as a unit to perform one or more functions. . . ." Thus we may think of initiating devices as the smoke detectors, heat detectors, pull stations, or sprinkler valve supervisory switches that send a signal to the fire alarm panel while using little or no energy themselves. Notification appliances include those items such as speakers, horns, sirens, and strobes that use energy to produce an audible or visual signal, or both, to notify the occupants of a building in case of a fire. Typically, notification appliances do use much more energy than initiating devices.

Non-Power-Limited Circuits

Non-power-limited fire alarm circuits are limited only by voltage to a maximum of 600 volts. The *NEC* also requires the power sources to meet the requirements of the first four chapters of the *Code*. In effect, NPLFA circuits are treated the same as branch circuits, although Article 760 does modify those general requirements in significant ways. One significant restriction is that NPLFA circuits may not be supplied through ground fault circuit interrupters (GFCIs) even where they are supplied through receptacles that would otherwise require GFCI protection, such as receptacles in unfinished basements of dwelling units. This rule is found in Section

760.21, and the corresponding exception for unfinished basements is found in Section 210.8(A)(5) Exception No. 3. The rule in Article 760 prohibiting GFCI protected circuits is intended to increase the reliability of the power supply to the fire alarm system. The exception that accommodates Article 760 permits the non-GFCI protected receptacle to supply *only* a permanently installed fire alarm system. In a dwelling unit unfinished basement, additional receptacles that are GFCI protected would also be required for use by people with cords and tools.

Power-Limited Circuits

Power-limited fire alarm circuits are also prohibited from being supplied by a GFCI protected circuit or receptacle, and in that respect alone they are the same as NPLFA sources. The power sources for PLFA circuits must be listed PLFA sources or listed Class 3 sources, or be derived from listed PLFA equipment. In fact, some listed equipment, such as household fire warning systems and combination burglar and fire alarm systems, may use integral or separate power supplies that are listed as Class 2 sources, like the panel shown in Figure 4.2. The permitted energy levels for listed PLFA sources are given in Tables 12(A) and 12(B) in Chapter 9 of the *NEC*. If we compare these tables to Tables 11(A) and 11(B), we can see that the power limitations for PLFA sources cover a range of values similar to the combination of Class 2 and Class 3 sources.

Like the tables for Class 2 and Class 3 sources, Tables 12(A) and 12(B) are for listing purposes and cannot be used to make a field-constructed power supply. The listing requirements go well beyond the power limitations, and many other design specifications and tests are implied by listing. Power supplies that are an integral part of a listed fire alarm panel are very common, and such power supplies are covered by the various product standards that cover different types of fire alarm control panels. Such control panels are marked (see Figure 4.3) to indicate if they supply circuits that are power-limited. Separate power sources must also be marked to indicate the class of power supply.

FIGURE 4.2 Household burglar alarm and fire warning panel.

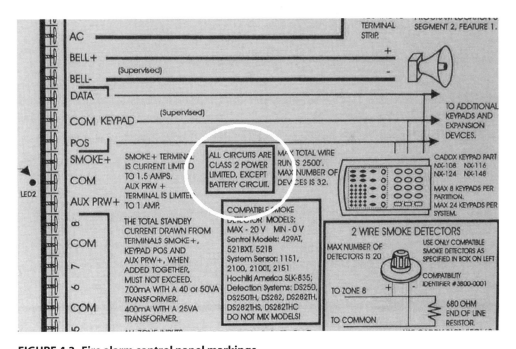

FIGURE 4.3 Fire alarm control panel markings.

GENERAL REQUIREMENTS FOR FIRE ALARM CIRCUITS

Article 760 modifies the ordinary requirements of the first four chapters of the *Code* in ways similar to Article 725. To begin with, Section 760.3 excludes fire alarm circuits from the requirements of Article 300 except where specific sections are referenced in Article 760. Then, in Section 760.3, specific sections, such as Section 300.21 stating that electrical installations may not substantially increase the spread of fire or products of combustion, are returned to apply to both NPLFA and PLFA circuits. The requirement that abandoned cables be removed so that they do not add unnecessarily to the fire load appears in Section 760.3(A). According to 760.3(B), Section 300.22, which covers wiring in ducts, plenums, and other spaces for environmental air, applies to fire alarm circuits, although it is modified somewhat for certain PLFA cables. Section 300.6 is also made applicable by Section 760.3(D) and requires fire alarm circuits to be installed in methods appropriate for the environment, especially with regard to corrosion protection.

Other general requirements that apply to all fire alarm circuits are the rules of Sections 760.5 and 760.6. These rules require that installations be done in a "neat and workmanlike" manner; that cables exposed on ceiling or wall surfaces be supported from structural elements of a building; that cables be protected from damage by screws or nails where they run parallel to framing members; and that cables be installed in a way that will not prevent removal of panels that afford access to other equipment. Figure 4.4 shows an example of an installation that not only obstructs access, but also uses the grid as a support means, neither of which conditions are permitted by the *NEC*. For PLFA circuits that are often wired using special cable types without raceways, Section 760.57 prohibits the PLFA conductors from being supported by attaching the cables to any conduits or raceways. A similar requirement also applies indirectly to NPLFA cables because of the *Code* requirement that cables be supported from structural components and the fact that raceways cannot be considered as structural components of a building.

Section 760.10 requires that all fire alarm circuits be clearly identified at terminals and junctions. This rule is intended to ensure that

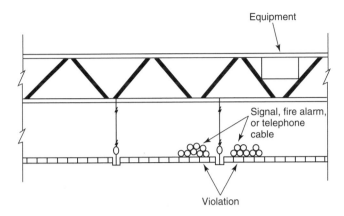

FIGURE 4.4 Misuse of grid ceiling.

the fire alarm circuits will not be unintentionally disturbed. There-fore, NPLFA and PLFA circuits must be marked, labeled, or tagged somehow so that the circuits can be readily recognized and not confused with other conductors and circuits. The differentiation is needed not only to distinguish fire alarm from power and lighting circuits, but also to distinguish between fire alarm circuits and other limited energy and signaling circuits. This identification is especially important where, for example, PLFA circuits have been reclassified and installed using NPLFA wiring methods, or where PLFA circuits have not been reclassified but are wired using NPLFA methods. Often, NPLFA circuits and reclassified PLFA circuits are wired using the same wiring methods as power and lighting circuits, and with-out identification they could easily be mistaken for branch circuits. Even where PLFA methods are used, cable types alone are not reli-able ways to identify circuits, especially where permitted cable sub-stitutions have been used. In some installations, the fire alarm system may have been installed using the same type of cable as the data and communications, so the wiring method alone is not at all reliable for identifying circuit classifications or uses. The requirement for the marking of PLFA circuits is specifically reiterated in Section 760.42 stating that PLFA equipment must be marked as such.

Section 760.7 applies to fire alarm circuits that extend beyond a single building. Such PLFA circuits either may comply with Parts II,

III, and IV of Article 800 or they are allowed to comply with Part I of Article 300. Parts II, III, and IV of Article 800 cover clearances and other installation requirements for outside conductors, requirements for protectors, and grounding of metallic cable sheaths and primary protectors. Part I of Article 300 covers all general wiring methods for conductors under 600 volts such as burial depths and physical protection requirements. Basically, PLFA conductors outside of a building may either be treated as communications conductors or as ordinary branch circuits.

Non-power-limited fire alarm circuits that run outside of a building also must meet the installation requirements of Part I of Article 300 and must meet the installation, clearance, and protection requirements of Part I of Article 225 that normally applies to outside branch circuits and feeders. Other requirements, such as the requirement for a disconnect where a branch circuit or feeder enters a building, do not apply to the fire alarm circuits because those requirements are in Part II of Article 225. In addition to addressing reliability of the fire alarm circuits, Section 760.7 addresses the hazards that may be created when a circuit leaves a building and is exposed to damage or contact with other circuits or exposed to lightning.

The requirements for circuits that extend outside of a building are illustrated in Figure 4.5. If the PLFA circuits in the diagram are treated as communications circuits and run where exposed to lightning as shown, communications circuit protectors would be required at both buildings.

Grounding of Fire Alarm Circuits and Equipment

Section 760.9 also applies to both NPLFA and PLFA circuits. It calls for fire alarm circuits and equipment to be grounded as required by Article 250. While this requirement may seem to apply the general rules of Article 250 to all the circuits covered by Article 760, Article 250 actually provides special rules for fire alarm circuits.

Article 250 requires enclosures and raceways of fire alarm circuits to be grounded if the system itself is required to be grounded,

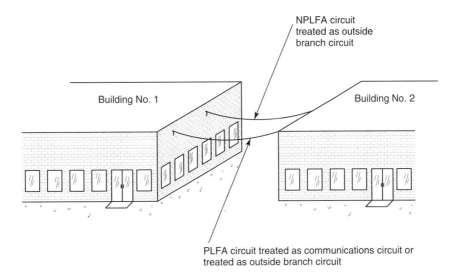

FIGURE 4.5 Outside fire alarm circuits.

that is, if the system is required to have a grounded conductor. On the one hand, most PLFA circuits are not required to be grounded, so any metal boxes, raceways, or other enclosures for those conductors are likewise not required to be grounded. On the other hand, *some* NPLFA circuits are required to be grounded, and if the circuit is required to be grounded, the metal raceways, boxes, and other enclosures are also required to be grounded.

The general approach of Article 250 is that any metal part that is likely to become energized due to the possible failure of insulation on an associated conductor should be grounded. Grounding will significantly reduce the possibility of a shock hazard. However, most PLFA circuits are below the voltage level that is considered to create a shock hazard. Therefore, even if the circuit did energize a conduit or other metal part that is subject to contact by persons, the shock hazard would still be minimal.

This discussion of grounding thus far summarizes and generalizes the rules, but we should examine the actual pathway through the *Code* that provides the rules. Section 250.86 says that "Except as permitted by 250.112(I), metal enclosures and raceways for other than service conductors shall be grounded." Three exceptions are provided—for existing installations without grounding conductors, for short sections of raceways used to protect cables, and for isolated buried underground elbows—but the primary "exception" is provided by Section 250.112(I). Section 250.112 requires grounding of exposed, non-current-carrying metal parts of specific types of equipment. (Remember, "equipment" is a broad term that includes virtually every item of material used in an electrical installation.) Section 250.112(I) says equipment supplied by fire alarm circuits is required to be grounded if the *system* is required to be grounded by Part II or Part VIII of Article 250. Part II covers alternating current systems and Part VIII covers direct current systems. The requirements for grounding a system are first based on voltage and then on other conditions. The actual requirement for grounding a system depends on the voltage of the initiating and notifying circuits of the fire alarm system.

Most new fire alarm systems are direct current systems that operate at about 24 volts. Many of their panels have PLFA power supplies for the low-current initiating device circuits (pull stations, detectors, and the like) and one or more NPLFA power supplies for the indicating appliance circuits (horns, speakers, strobes, and the like). Both types of circuits are likely to be 24 volt DC circuits, and we see in Section 250.162 that these voltage levels are not even mentioned and are not required to be grounded. Even DC fire alarm systems operating at over 50 volts up to 300 volts are excluded from the grounding requirements by Exception No. 3. (Exception No. 3 is really directed at an old-style fire alarm system or device, but according to Table 12(B), DC PLFA systems may be as high as 250 volts and 30 milliamperes.)

Since the system is not required to be grounded, the raceways, boxes, and other enclosures are not required to be grounded. Note, however, that the fire alarm control panel is likely to be supplied by a 120 volt AC circuit, and Section 250.20(B) does require the 120 volt system to be grounded, so the panel must also be grounded (see Figure 4.6). Some metal raceways containing the derived NPLFA or PLFA circuits may be grounded by their connection to the panel, but the grounding of those raceways, in this common example, is permitted but not required.

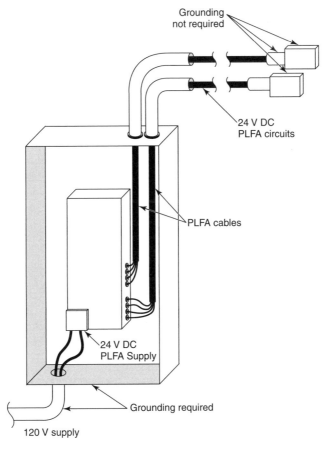

Grounding not required

24 V DC PLFA circuits

PLFA cables

24 V DC PLFA Supply

Grounding required

120 V supply

FIGURE 4.6 Fire alarm equipment grounding.

Whether a fire alarm system is required to be grounded depends on the characteristics of that system. Consequently, the conclusion from case to case may be different, but the path through the *Code* requirements is essentially the same.

Some people ask why anyone would want to leave a box or a raceway ungrounded. First, as already mentioned, even if the 24 volt DC system did energize the conduit, it would not usually represent a shock hazard or fire hazard. Second, a box may be required for a manual pull station or some other initiating device or notifying appliance. Where cable wiring methods are being used, the cable usually does not have an extra conductor for grounding the box or a section of raceway that affords access to the box for a cable. Third, many fire alarm power systems need to be ungrounded to improve reliability and function. The use of an ungrounded system can allow some panels to experience a single ground on a circuit, detect the damage, send a trouble signal, and continue to function.

Equipment Requirements and Power Supplies

Section 760.41 specifies that PLFA power sources are required to be listed. In some cases, the power sources are separate items and may be separately listed. In most cases, the power supplies are part of listed fire alarm control panels or may take the form of a module that is listed for use with a control panel. Most wiring methods in the *NEC* are also required to be listed by the individual articles in its Chapter 3. Sections 760.31 and 760.71 state that NPLFA and PLFA cables are required to be listed.

The *NEC* does not cover the panels, devices, and appliances that are used with fire alarm circuits. The *National Fire Alarm Code* (*NFPA 72*) requires all equipment constructed and installed in conformity with *NFPA 72* to be "listed for the purpose for which it is used." This rule is found in Section 1-5.1.2 of *NFPA 72*. The following is quoted from the commentary to Section 1-5.1.2 in the *National Fire Alarm Code®️ Handbook:*

> Fire alarm products must be listed for the specific fire alarm system applications for which they are used. Because fire alarm systems are used for life safety, property protection, mission continuity, heritage preservation, and environmental protection, the listing requirements are often more stringent than for those products listed for electrical safety only.
>
> Equipment listings generally contain information pertaining to the permitted use, required ambient conditions in the installed location, mounting orientation, voltage tolerances, compatibility,

and so on. Equipment must be installed in conformance with the listing to meet the requirements of the code. [NFPA, *National Fire Alarm Code Handbook*, 1999, p. 361]

The *NEC*, in Section 110.3(B), also requires that all listed or labeled equipment be installed in accordance with the instructions included in the listing or labeling. This requirement seems almost self-evident, but reading and following instructions do not come naturally to everyone. Nevertheless, one point made by the preceding commentary is that only by following the instructions can equipment be expected to fulfill its required functions. Thus, installers and inspectors should look for the listing and labeling instructions on equipment and be certain that those instructions are followed as indicated. For example, Figure 4.7 shows a pull station.

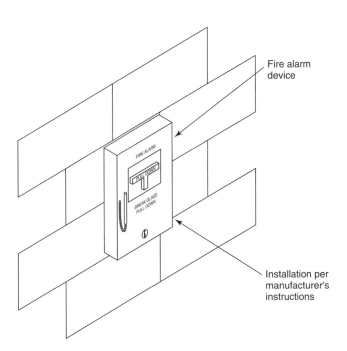

FIGURE 4.7 Manufacturer's instructions.

The manufacturer's instructions for this pull station may include detailed information such as the required size and type of back boxes.

APPLICATION OF NON-POWER-LIMITED CIRCUITS

Non-power-limited fire alarm circuits are widely used in modern fire alarm systems, mostly for powering notifying appliance circuits. Some notifying appliances, such as the strobe light illustrated in Figure 4.8, use a significant amount of power to operate. Most smoke detectors and similar initiating devices use very little power in their operation, and some devices, such as manual pull stations, use no power at all. In some cases more than one NPLFA power supply may be needed to power a group of horns, strobes, or speakers. In most modern applications, NPLFA power sources are still only

FIGURE 4.8 Typical visible notification appliance. (Source: *National Fire Alarm Code® Handbook,* NFPA, 1999, Exhibit 1.24; courtesy of Gentex Corp.)

24 volts or so, but the energy they supply takes them outside the limits of PLFA or Class 3 power supplies. While NPLFA sources are permitted to be up to 600 volts and some equipment controlled by fire alarm panels may require control voltages of 120 volts or more, such circuits are not considered to be fire alarm circuits unless they are "powered and controlled" by a fire alarm system. Circuits that are only controlled by a fire alarm system, such as fan shutdown circuits, are remote-control or signaling circuits and are covered by Article 725. Figure 4.9 illustrates the difference between fire alarm circuits and power circuits or circuits covered by Article 725.

In many respects, NPLFA circuits are treated the same as Class 1 circuits, but the rules for NPLFA and Class 1 circuits are not identical. Article 760 does modify the ordinary rules of the first four chapters of the *NEC* in very significant ways that are much like the modifications found in Article 725 for Class 1 circuits. Unlike Article 725, Article 760 does not contain a Fine Print Note that summarizes the ways that the general rules are modified. Nevertheless, the numbering and arrangement of Articles 725 and 760 are quite similar and so are the ways that the general rules are changed. In the

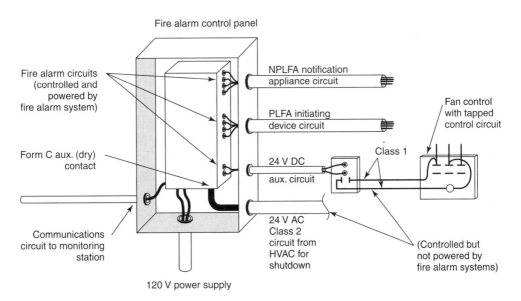

FIGURE 4.9 Fire alarm and other circuits.

following sections we look at how Article 760 changes the rules with regard to minimum wire sizes, derating factors, overcurrent protection, insulation requirements, wiring methods and materials, and circuit separations.

Minimum NPLFA Wire Sizes

The minimum size of conductors for branch circuits, 14 AWG copper, is given in Section 310.5. However, Section 310.5 Exception No. 7 recognizes the rules of Article 760 for conductor sizes in fire alarm circuits. Of course, Article 725 can modify the requirements of Article 310 without "permission" from Article 310, but such an exception often helps avoid misinterpretations.

Section 760.27(A) provides the special rules for the sizing of NPLFA circuit conductors. This rule permits the use of 16 AWG and 18 AWG copper. Although this section does not specify copper, Section 110.5 says that "Where the conductor material is not specified, the material and sizes given in this *Code* shall apply to copper conductors." The use of 16 AWG and 18 AWG conductors is acceptable as long as the loads are limited to the ampacities given in Section 402.5. The ampacities for larger wires are given in Section 310.15. Most of the types of wires listed in Article 310 are not available in sizes smaller than 14. Note that the ampacity given in Section 402.5 limits the load on the wires; for an 18 AWG conductor the load is limited to 6 amperes, and for 16 AWG the load is limited to 8 amperes.

Article 402 covers fixture wires. Normally, fixture wires are permitted to be used only as part of luminaires (lighting fixtures) or similar equipment, or for connecting to luminaires. Fixture wires are not permitted as branch-circuit conductors. However, Section 760.27(B) modifies these requirements somewhat by permitting certain types of fixture wires to be used for NPLFA circuit conductors. Other types are also permitted if listed for NPLFA use. Two types of fixture wire that are often used for NPLFA circuits are Type TFN and Type TFFN. Table 402.3 describes these wire types. They are essentially equivalent to Type THHN as both are thermoplastic insulations, usually with nylon jackets.

Section 760.27(C) requires conductors to be copper and permits either solid or stranded conductors. In fact, some of the permitted types of fixture wires in Section 760.27(B) are only stranded types while others are usually solid. For example, Type TFFN is only a flexible stranded conductor (probably 19-stranded), while Type TFN may be either solid or 7-stranded. The fact that Type TFFN is a flexible stranded conductor should not be taken to mean that the wire is intended for applications where flexibility is required in normal use. According to Section 402.10, fixture wire is not intended for use where it is subject to bending and twisting in use.

One concern with stranded wire is that a single strand may make contact at a terminal and seem to be in working order but be incapable of actually carrying current in an alarm condition. For this reason, prior to the 1996 *NEC*, Article 760 required solid or bunch-tinned conductors generally and permitted stranded wires only if they had no more than a certain number of strands, with the number of strands depending on the size of the wire. This restriction was removed in the 1996 *NEC*.

Section 760.27(C) Exception allows two other types of conductor insulation that are not listed in subsection 760.27(B), but these types, Types PAF and PTF, may only be used in high-temperature applications. The Exception applies to both subsections 760.27(B) and 760.27(C) because it permits types not listed in 760.27(B) and permits conductors that are nickel or nickel-coated copper rather than just copper as required by 760.27(C).

As was noted earlier, NPLFA circuits are often used to power relatively high-current notifying appliances such as strobes. These notifying appliances have markings that indicate the minimum voltage at which they will still be bright enough (for strobes) or loud enough (for horns or speakers) to meet the requirements of the *National Fire Alarm Code*. This need for a certain level of light and sound makes voltage drop a critical consideration on such circuits, and even where the ampacity of a conductor is sufficient for the load, the conductor size may be required to be larger than the minimum size in order to ensure adequate voltage at the most distant appliance. In most applications of the *NEC*, there are no maximum voltage drop rules because voltage drop due to undersized wires or long runs is primarily a performance issue and not a safety issue. In the case of fire alarm circuits and a few other cases, such as fire pump circuits, performance is also a definite safety issue

Derating Factors in NPLFA Circuits

"Derating factors" generally refers to the temperature correction factors that appear at the bottom of the allowable ampacity tables, and to the adjustment factors for more than three conductors in a raceway or cable that appear in Section 310.15(B)(2)(a). The derating factors of Section 310.15(B)(2)(a) apply only to current-carrying conductors, and only when there are more than three current-carrying conductors in a raceway or cable. The question is, what conductors count as current carrying?

Section 760.28 says, in effect, that only NPLFA circuit conductors that are loaded to more than 10 percent of their ampacities count as current-carrying conductors. Also, where NPLFA conductors share a raceway with Class 1 conductors, as permitted in Section 760.26, only those Class 1 or NPLFA conductors that are loaded to over 10 percent of their ampacities are counted as current-carrying conductors for the purpose of applying Section 310.15(B)(2)(a) adjustment factors. The application of derating factors to power conductors is not changed by this rule. Power conductors count as current-carrying conductors, but derating factors do not apply until there are a total of more than three power conductors and other conductors that also count as current carrying. Section 760.26 permits NPLFA conductors to share a raceway or other enclosure with power conductors if both types of conductors connect to the same equipment. In any case, the derating factors do not apply until there are more than three current-carrying conductors in a raceway or cable.

Conduit fill limits apply to NPLFA conductors and to any power conductors or Class 1 conductors that are also in the same raceway.

Note that the only derating factors actually referenced in Section 760.28 are the "adjustment factors" and not the "correction factors" used for temperature adjustment. Ambient temperatures and operating temperatures of associated conductors must still be considered. Section 402.5 states that "No conductor shall be used under such conditions that its operating temperature exceeds the temperature specified in Table 402.3 for the type of insulation involved." A Fine Print Note then references Section 310.10 that provides further information on the temperature limitations of conductors.

Section 402.5 does not say how temperature issues should be considered or how ampacities should be adjusted. However, Table 402.3 does provide temperature ratings for conductor types. The types mentioned earlier, TFN and TFFN, are both 90°C conductors. Although the *NEC* never specifically says so, the correction factors from the 90°C column of Table 310.16 or Table 310.17 (the factors are the same) should also be usable for temperature corrections for TFN or TFFN fixture wires.

Another significant temperature consideration in selecting NPLFA circuit conductors is that conductors cannot be associated together in such a way that the operating temperature of any conductor is exceeded. Section 402.5 and Section 310.10 both refer to this issue. For example, suppose we select from Section 760.27(B) a Type TF or TFF for our NPLFA circuit conductors. From Table 402.3 we find that these are both 60°C conductor types. If we were to mix these conductors with Type THHN Class 1 or power conductors in a raceway as permitted by Sections 760.26, the THHN conductors could not be used at their 90°C ampacities, or even at the lower 75°C ampacities. The THHN conductors would be limited to 60°C ampacities to avoid overheating of the 60°C NPLFA conductors. This example is illustrated in Figure 4.10.

NPLFA Overcurrent Protection

Section 760.23 covers overcurrent protection for NPLFA circuit conductors. This section says the conductor ampacity may be used and the derating factors of Section 310.15 may be disregarded in determining the ampacity for the purposes of selecting overcurrent protection. This rule applies only to conductors 14 AWG or larger. Overcurrent protection for smaller conductors is specified to not exceed 7 amperes for 18 AWG and 10 amperes for 16 AWG. Thus,

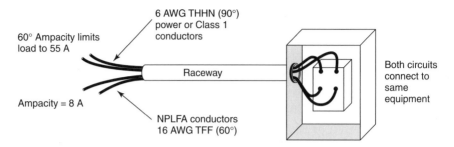

FIGURE 4.10 Conductor temperature limitations.

for all conductors under this rule, the maximum values for over-current protection are or may be somewhat higher than the ampacities of the wires. This provision should not be surprising if the nature of the loads on NPLFA conductors is considered.

Most NPLFA are not so heavily loaded or loaded for so long that overheating is likely. As previously noted, because of voltage drop considerations, those NPLFA conductors that are heavily loaded under alarm conditions are likely to be oversized to compensate for voltage drop anyway. The most likely overcurrents are short circuits and ground faults. In other areas of the *NEC,* where overloading is a significant risk, overcurrent protection is usually not permitted to exceed the ampacity of a conductor. In fire alarm circuits, over-loading is not likely in most cases. In fact, the Exception to Section 760.23 recognizes other rules of the *Code* that may permit or require other levels of overcurrent protection.

In general, overcurrent protection for NPLFA circuits should be located where the conductors receive their supply. However, Section 760.24 permits alternative locations in much the same manner as the tap rules of Section 240.21, and the rules for NPLFA conductors differ a bit from the ordinary feeder tap rules. Section 760.24, Exception No. 1 permits an NPLFA conductor to be tapped from a larger conductor where the overcurrent device for the larger conductor is sized to also protect the smaller tap conductor. As shown in Figure 4.11, this rule may be applied where a conductor is reduced in size near the end of a circuit that has been oversized for voltage drop correction.

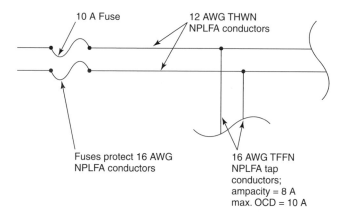

10 A Fuse

12 AWG THWN
NPLFA conductors

Fuses protect 16 AWG
NPLFA conductors

16 AWG TFFN
NPLFA tap
conductors;
ampacity = 8 A
max. OCD = 10 A

FIGURE 4.11 NPLFA overcurrent device location and rating.

Section 760.24, Exceptions No. 2 and No. 3 permit NPLFA conductors to be protected by the overcurrent device on the supply side of a single-phase transformer with a two-wire secondary or a similar listed electronic power supply with a two-wire output. In both of these rules, the ampacity of the NPLFA circuit conductors on the secondary or output is multiplied by the secondary-to-primary or output-to-input voltage ratio, and the result must not be less than the rating of the primary overcurrent device. The primary overcurrent device must also protect the transformer in accordance with Section 450.3. See Figure 4.12 for an example of the secondary conductors of a transformer being protected by a primary overcurrent device.

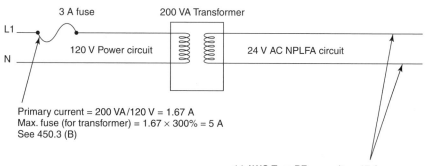

3 A fuse

200 VA Transformer

L1

120 V Power circuit

24 V AC NPLFA circuit

N

Primary current = 200 VA/120 V = 1.67 A
Max. fuse (for transformer) = 1.67 × 300% = 5 A
See 450.3 (B)

14 AWG Type PF ampacity = 17 A
Secondary/Primary voltage ratio = 24/120 = 0.2
Max. fuse (for NPLFA conductors) = 17 A × 0.2 = 3.4 A
3 A Primary fuse protects 17 A conductors

FIGURE 4.12 NPLFA secondary conductor overcurrent location and rating.

NPLFA Insulation Requirements

Insulation requirements are not significantly altered for NPLFA circuits. Section 760.27(B) requires NPLFA circuit conductors to be suitable for 600 volts. The thermoset and thermoplastic types listed in Table 310.13 and permitted for branch circuits or feeders are all 600 volt or higher rated insulation types. Although Table 402.3 does include some 300 volt rated insulation for some fixture wires, none of the 300 volt types are listed in Section 760.27(B) as suitable for NPLFA wiring.

Section 760.30 permits special multiconductor cables that are listed as NPLFA cables to be used for fire alarm circuits of up to 150 volts. However, multiconductor NPLFA cables also are required to be rated for 600 volts according to Section 760.31(B), but are only permitted to be marked with the maximum usage voltage rating of 150 volts. The application of these cables is covered in the following section on wiring methods and materials.

Wiring Methods and Materials for NPLFA Circuits

The previous sections of this chapter have shown how NPLFA circuits are treated differently from branch circuits with regard to minimum wire sizes and types of wire, the use of derating factors, and overcurrent protection. Earlier in this chapter it was stated that Article 760 excludes fire alarm circuits from most of the requirements of Article 300, but that certain provisions of Article 300 are specifically referenced, such as the requirements for wiring in ducts, plenums, and other air-handling spaces; the requirements to not increase the spread of fire; and certain requirements that apply when fire alarm circuits extend outside of a building. However, Section 760.25 also requires that other sections of Article 300 apply to NPLFA circuits, specifically, Sections 300.11(A), 300.15, and 300.17. Section 110.3(B) that requires compliance with listing and labeling instructions is also mentioned, but this reference serves more as a reminder than as a reintroduced requirement.

The other specific references do reintroduce requirements from Article 300. Section 300.11(A) requires secure support of wiring methods and prohibits the use of a ceiling grid for support of cables or raceways. Section 300.15 requires that boxes be provided for all splices, outlets, junctions, terminations, or pull points for NPLFA raceways and cables. And, as also mentioned in Section 760.28, the conduit fill restrictions of Table 1 in Chapter 9 must be observed in accordance with Section 300.17. Whatever articles in Chapter 3 of the *Code* are applicable to the particular wiring method chosen must apply as well.

Based on the main rule in Section 760.25, there would seem to be no special wiring methods stipulated for NPLFA wiring. Users of NPLFA circuits can choose from the ordinary wiring methods of Chapter 3 of the *Code*. The general rules of Chapter 3 are modified as previously explained, and these modifications are recognized by Section 760.25 Exception No. 1. Exception No. 2 to Section 760.25 incorporates any restrictions that may be imposed by other articles, such as the restrictions on wiring methods in hazardous areas and other specific occupancies found in Chapter 5 of the *Code*. However, in addition, Exception No. 1 recognizes the special purpose multiconductor NPLFA cables permitted by Section 760.30. These cables are unique to NPLFA circuits. Although in most respects NPLFA circuits are treated the same as Class 1 circuits, the special cables and the rules about circuit separations are significant differences between NPLFA and Class 1 circuits.

The permitted use of NPLFA cables provides an obvious installation advantage where the cable can be installed without a raceway. However, the uses of NPLFA are significantly restricted. Essentially, NPLFA cables can be run concealed in most areas without raceways, but where exposed, the cables must be protected from damage by raceways or other means. Also, NPLFA cables must be protected by raceways where they run through floors or are installed in hoistways. These requirements can be found in Section 760.30(A).

Section 760.30(B) provides requirements for the application of specific listed NPLFA cable types in specific types of locations. Listed NPLFA cables are available in Types NPLFP, NPLFR, and NPLF that stand for non-power-limited fire cable, with the additional *P* signifying "plenum" and the additional *R* signifying "riser." However, none of these types are permitted to be run exposed in a duct or plenum according to Section 760.30(B)(1). In such spaces, any of the

NPLF cable types must be installed in a raceway, and if the Fine Print Note under Section 760.30(B)(1) is followed to Section 300.22(B), such wiring is permitted in ducts and plenums for environmental air only where the wiring connects to equipment that acts directly on the contained air. Since Section 760.3(B) refers to Section 300.22 generally, Section 300.22(A) prohibits any such wiring from being installed in ducts that handle loose stock or vapor, with or without raceways. "Other spaces for environmental air" are covered in Section 760.30(B)(2), and in these spaces Type NPLFP may be used without raceways, or Types NPLFR or NPLF can be used in metallic raceways.

A suspended air-handling ceiling is the most common example of an "other space for environmental air." These ceilings are commonly called "plenum ceilings" in the field, and could be construed as plenums under the definition of a plenum in Article 100. However, the *NEC* makes a distinction between plenums that are specifically fabricated and constructed primarily to handle air and other spaces used for handling environmental air but not constructed primarily and exclusively for that purpose. Section 300.22 must be read in its entirety and carefully in order to understand the subtle distinctions being made between ducts and plenums, and other space for environmental air.

Riser applications include vertical installations that pass through more than one floor and vertical runs in a shaft. Such applications require the use of NPLFR cable. Other NPLF cables may be used if installed in raceways or installed in fireproof shafts with firestops at each floor.

Type NPLF, NPLFR, or NPLFP cables may be used in other spaces in buildings that are not in other space for environmental air or in risers.

As noted, the *P* and *R* suffixes may be used to indicate that an NPLF cable is suitable for plenum or riser use. Plenum cables must not spread a fire rapidly or produce a lot of smoke that would be transported through the air-handling system of a building. Riser cables must pass a riser flame test to ensure that the cable is slow in propagating flames so that a fire will not spread from floor to floor by way of the cable.

Another suffix that may be used is *CI,* which stands for "circuit integrity." The *CI* suffix may be used in addition to the other markings on a cable, so cables may be designated NPLF-CI, NPLFR-CI, or NPLFP-CI. Circuit integrity cable is designed to be able to maintain its function (integrity) for a certain amount of time in a fire. *NFPA 72* requires that certain circuits be arranged and installed to extend their survivability in a fire. Type CI cable provides an alternative to other protection methods. The other protection methods depend on whether the circuit is an initiating circuit, a notification circuit, or a circuit between a central control panel and a remote fire command center. Type CI cable is required to have a 2-hour fire rating. Figure 4.13 shows an example of a circuit integrity cable. The only other cable type that has a similar rating is Type MI (mineral insulated) cable (see Figure 4.14). The requirements for survivability are beyond the scope of this book, but the various fire alarm circuit classes and styles are covered briefly under the heading "Related Requirements" near the end of this chapter.

NPLFA Circuit Separations

Although for NPLFA circuits we may choose from the same wiring methods used for branch circuits, Article 760 imposes some additional separation requirements on installations of NPLFA circuit conductors, including those installed using multiconductor NPLFA

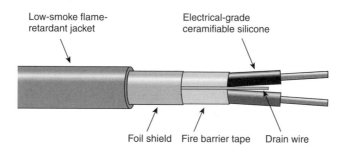

FIGURE 4.13 Typical circuit integrity (CI) cable. (Source: *National Electrical Code®* **Handbook,** NFPA, 2002, Exhibit 760.5; courtesy of Rockbestos Surprenant Cable Corp., Clinton, MA)**

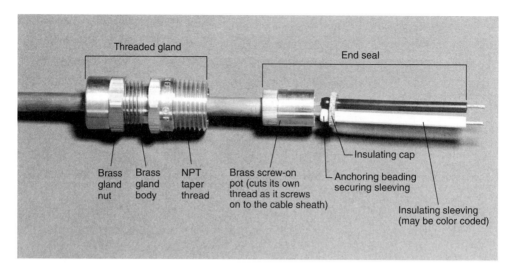

FIGURE 4.14 Typical mineral insulated (MI) cable. (Source: *National Electrical Code®* *Handbook,* 2002, Exhibit 501.4; courtesy of Pytrotenax USA, Inc.)

cables. However, for NPLFA circuits, the separation requirements are not nearly as stringent as they are for Class 1 circuits. Section 760.26(A) permits mixing of NPLFA and Class 1 circuits without restriction and without a requirement that they be functionally associated or connect to the same equipment. Actually, there is one restriction: All the conductors of both types of circuits must be insulated for the maximum voltage of any contained circuit. However, both Class 1 and NPLFA circuits are limited to 600 volts and are required to use 600 volt insulation, so this restriction is easily met if the other requirements are met. This permission to mix NPLFA and Class 1 circuits does not conflict with Article 725 because that article does not mention any restriction on such use. Section 725.55 does prohibit mixing of Class 2 or Class 3 circuits with NPLFA circuits.

Non-power-limited fire alarm circuits are also permitted to occupy the same cable, enclosure, or raceway with power circuits, but only where both types of circuits connect to the same equipment. This restriction differs somewhat from the rule in Article 725 that allows mixing Class 1 and power conductors only where they are functionally associated. In most cases, circuits that are functionally associated will also connect to the same equipment, but the language in Section 760.26(B) is different and, perhaps, slightly more restrictive. For example, NPLFA circuits may be used to control other equipment, such as a fan, and will connect to the same equipment as the fan circuit at the motor control. However, a pressure switch that proves operation of the fan could be said to be functionally associated, but does not necessarily connect to the same equipment. This language difference is not coincidental. The language was identical in the 1975 *NEC* when Article 760 first appeared in the *Code*, but in 1978, Article 725 changed to require only functional association, not connection to the same equipment, while the language in Article 760 has remained unchanged and requires connection to the same equipment.

These separation rules, found in Section 760.26, are different from the general rules found in Article 300. In Article 300, conductors of different circuits may share a common enclosure if they are all 600 volts or less, and if all conductors are insulated for the maximum circuit voltage in the enclosure. Generally, conductors of circuits that are over 600 volts cannot be mixed with circuits that are 600 volts or less. Article 760 is significantly more restrictive than Section 300.5(C)(1), as well as being somewhat more restrictive than Article 725.

APPLICATION OF POWER-LIMITED CIRCUITS

Power-limited fire alarm circuits are represented by most of the initiating device circuits in modern fire alarm systems. In some systems, such as household fire warning systems, the notification appliances are also powered by PLFA circuits. Some other ancillary equipment, such as auxiliary relays for controlling fans or fire dampers, may also be powered and controlled from a fire alarm control panel through PLFA circuits.

Power-limited fire alarm circuits are defined by their power supplies. The power supplies limit the total energy in a PLFA circuit to a defined maximum value that will not be exceeded even with a short circuit on the load side of the power supply. These inherent energy limitations are what make PLFA circuits limited energy or power-limited circuits. The actual permitted values of voltage and current that are permitted to be delivered by PLFA power sources are approximately equivalent to Class 3 sources. All PLFA sources must be listed and marked to indicate the class of supply and the electrical rating, and they may be marked as PLFA or Class 3 sources. In fact, in some cases, fire alarm panels may be listed with Class 2 power supplies.

Generally, the classification of a power supply can be readily determined from the marking on the power supply, or if the power supply is part of other listed equipment, from the marking on that equipment or at the limited energy terminals of the equipment. Once the classification of the power supply has been determined and verified, how are the circuits to be treated? As shown for NPLFA circuits, Article 760 modifies the requirements of Chapters 1 through 4 of the *NEC* with regard to five primary issues: minimum wire sizes, derating factors, overcurrent protection, insulation requirements, and wiring methods and materials. We now examine each of these issues as they relate to PLFA circuits.

Minimum PLFA Wire Sizes

Installers and designers of PLFA circuits have three choices of wiring methods. First, the circuits can be installed using NPLFA wiring methods. To use NPLFA methods, the PLFA circuits may be reclassified as NPLFA circuits or remain classified as PLFA circuits. Second, the circuits can be installed using cable types specifically listed for PLFA use. Third, the PLFA circuits can be installed using substitute listed cable types.

If the circuits use NPLFA methods, they must comply with Section 760.25, which recognizes the requirements of Sections 760.26 through 760.30, so the minimum wire size will be the same as for NPLFA circuits, that is, 18 AWG. This requirement is reinforced by Section 760.58 that requires single conductors to be 18 AWG or larger. Otherwise, the *NEC* specifies listed cables, and Section 760.71(B) requires the conductors in listed cables to be no smaller than 26 AWG. Since the conductors must be terminated somewhere, the terminals used will also determine conductor sizes in some cases, because Section 760.58 permits 26 AWG conductors only where the terminals or wire connectors are listed as suitable for 26 AWG conductors.

Prior to the 1996 *NEC*, small conductors had to be a part of a cable with a certain minimum number of conductors. For example, cables composed of 26 AWG conductors had to have at least 10 conductors in the cable. These minimum numbers were based on concerns about tensile strength and pulling tension. However, according to UL 1424, *Standard for Safety for Cables for Power-Limited Fire-Alarm Circuits,* cables listed for PLFA circuits with fewer than 4 conductors and conductors smaller than 24 AWG (25 and 26) must pass a breaking strength test. Since the testing satisfies the concern, the rule about number of conductors in cables was deleted.

Typically, selection of conductor sizes is based on performance requirements or manufacturer's specifications. For a given application, a manufacturer of a fire alarm control panel will usually specify a minimum size for up to some maximum circuit length or for some maximum number of devices. These limits may be absolute limits per circuit or may be flexible to allow for more devices or longer lengths with larger conductors. Overcurrent protection ratings are not used to size PLFA conductors because the power supplies are inherently power-limited and may not even have overcurrent protective devices. Furthermore, the *NEC* does not provide ampacities for conductors smaller than 18 AWG.

Section 760.41 prohibits a PLFA power source from being supplied through a GFCI device, and Section 760.51 establishes requirements for the wiring on the supply side of a PLFA power source. This section requires that the overcurrent protection ahead of the power source be no greater than 20 amperes. However, the exception to this section permits the input leads of a transformer to be smaller than 14 AWG if the leads are not longer than 12 inches. The minimum size for these transformer leads is 18 AWG and the conductor must be of a type permitted for NPLFA circuits. The exception does not change the overcurrent device rating; it only permits smaller wires to be connected to the maximum 20 ampere overcurrent device as shown in Figure 4.15. This permission is very similar to what is permitted for listed appliances in which the load is known and over-loading is not a significant risk. For example, Sections 240.4(A) and 240.4(B) permit 18 AWG cord on a listed appliance or 18 AWG fixture wire to be supplied from a 20 ampere branch circuit.

Derating Factors in PLFA Circuits

Article 760 does not say anything about derating of PLFA circuit conductors installed as PLFA circuits. Nor does Article 310 provide ampacities for the conductor sizes most commonly used for these circuits. This omission is not an oversight. There is really no need for such rules since the power supplies are inherently energy limited even under short-circuit conditions. Article 760 permits the use of small conductors because there is little risk of overloading, power supplies are limited, and the levels of power limitation keep the cir-

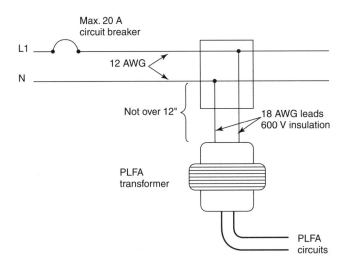

FIGURE 4.15 Overcurrent protection on the supply side of a PLFA source.

cuits from being fire hazards even if the conductors were to become overloaded, damaged, or otherwise faulted. When NPLFA methods are used for PLFA circuits, Section 760.52 (A) Exception No. 1 says that "the derating factors given in 310.15(B)(2)(a) shall not apply."

When the term *derating factors* is used with regard to the conductor ampacities of Article 310, it refers to "correction factors" applied to higher than normal ambient temperatures as well as "adjustment factors" applied to more than three current-carrying conductors. [The derating factors of Section 310.15(B)(2)(a) that are specifically mentioned are the "adjustment factors."] The low current levels that are typical in PLFA circuits make the conductors similar in heating characteristics to NPLFA conductors that are similarly loaded. Any NPLFA conductors that carry loads not over 10 percent of their ampacities are not required to be counted as current-carrying conductors. Such conductors are simply not adding significant heat to the installation. Nevertheless, all conductors and cables must be applied within their ratings.

Power-limited fire alarm cables do have temperature ratings. These temperature ratings are required to be marked on the cables if they are other than 60°C. Many cables have no temperature markings, and where there are no markings, and the cable is listed, the cable can be assumed to be rated for 60°C. Cables with higher temperature ratings are common, especially up to 90°C and will be marked as such. Depending on the properties of the insulation and jacket materials, the temperature rating of a PLFA cable may be as high as 250°C. (This temperature limit does not consider circuit integrity cables.) Section 110.3(B) says that listed equipment must be installed in accordance with the listing and labeling, and this rule applies to the cable types that are listed for use with PLFA circuits.

PLFA Overcurrent Protection

Overcurrent protection is inherent in the power supplies for PLFA circuits. The only significant rule about separate overcurrent protection for these circuits applies solely to the supply side of a power supply. The overcurrent device on the supply side is limited to

20 amperes. This rule, found in Section 725.51, is also mentioned in the preceding discussion of minimum PLFA wire sizes.

Considering overcurrent protection does bring up a point about the use of NPLFA methods as opposed to the reclassification of PLFA circuits as NPLFA circuits. In both cases NPLFA methods will be used. When PLFA circuits are wired using NPLFA methods, the circuits remain as PLFA circuits. In fact, parts of a circuit, such as a part that goes outside and is buried underground, may use an ordinary wiring method from Chapter 3 of the *Code* that is suitable for direct burial in a wet location, whereas parts of the circuit that are located indoors may use PLFA cables as shown in Figure 4.16. Terminals in the panel must be clearly marked so that it can be identified as part of a PLFA circuit. However, when a PLFA circuit is reclassified as NPLFA, it *becomes* an NPLFA circuit, and NPLFA cir-

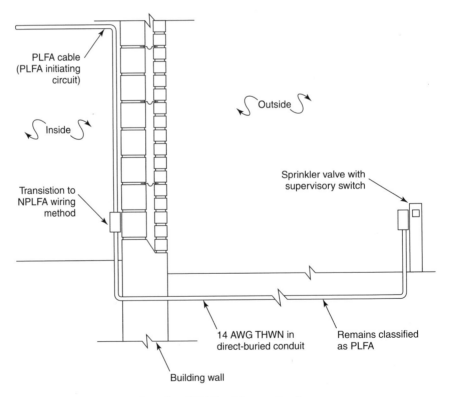

FIGURE 4.16 PLFA circuits using NPLFA wiring methods.

cuits are required to have overcurrent protection in accordance with Section 760.23.

There may be advantages to either approach. On the one hand, for example, if a circuit is reclassified and becomes an NPLFA circuit, it may share a raceway with Class 1 conductors, but it is required to be wired according to NPLFA rules everywhere. On the other hand, if a portion of a circuit runs through an area for which the PLFA cables are not suitable, the wiring method can be changed to NPLFA, but because it is still a PLFA circuit, all separation rules and other requirements for PLFA circuits continue to apply.

PLFA Insulation Requirements

The required insulation ratings for PLFA cables are found in Section 760.71(C). This section requires PLFA cables to have insulation rated at not less than 300 volts. All NPLFA circuits must have 600 volt insulation. However, unlike NPLFA wires and cables, PLFA are not permitted to have the voltage ratings or usage ratings marked on the cables. Cables that have listings in addition to the PLFA listings are permitted to have voltage markings if one of the additional listings requires the markings. Voltage markings on power-limited cables may be misconstrued to imply suitability for higher power uses, such as for Class 1 circuits or power or lighting circuits.

Wiring Methods and Materials for PLFA Circuits

As stated in the discussion of minimum wire sizes, there are three choices for wiring methods for PLFA circuits. The circuits can use NPLFA methods whether or not they are reclassified as NPLFA circuits; they can be installed using cable types specifically listed for the PLFA uses, or they can be installed using substitute listed cable types.

The main rule of Section 760.52 permits either the use of NPLFA wiring methods under subsection 760.52(A) or PLFA wiring methods under subsection 760.52(B). Section 760.52(A) permits the use of

NPLFA wiring methods without mention of reclassifying the circuit. The circuit would have to continue to be clearly marked as a PLFA circuit in accordance with Section 760.42. If a PLFA circuit is installed entirely or partially using NPLFA wiring methods, it remains a PLFA circuit. However, if a PLFA circuit is *reclassified* for some reason, it becomes an NPLFA circuit. Circuits that are reclassified become NPLFA circuits throughout their entire length as illustrated in Figure 4.17, even though they continue to be supplied by a PLFA power source.

Section 760.52 must be applied carefully. Section 760.52(A) Exception No. 3 contains permissive language as described in Section 90.5(B) and simply permits reclassification. Subsection 760.52(A) permits the use of NPLFA wiring methods without modifying the requirement for marking. Exception No. 3 permits the reclassification of the circuit, which then forces the use of NPLFA methods and requires the removal of the PLFA circuit markings. There is a difference between allowing the use of NPLFA wiring methods and

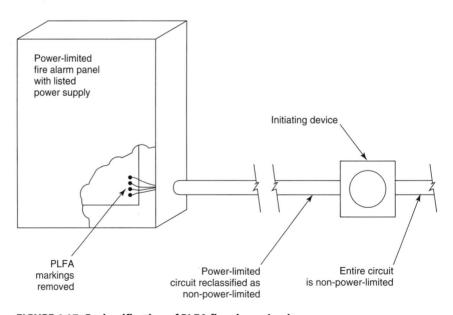

FIGURE 4.17 Reclassification of PLFA fire alarm circuits.

making the circuit an NPLFA circuit, and both possibilities are recognized in Section 760.52.

Perhaps the most popular choice for PLFA wiring methods comes from Section 760.52(B). This section permits the use of special wiring types that are not generally permitted for other uses. Thus, if we have a PLFA circuit, we can pick a PLFA cable type that is suitable for the occupancy and location within the occupancy. Such cables can generally be used without raceways if properly selected.

Section 760.71 describes the applications for which specific cable types are listed. The basic cable type is FPL, which stands for fire power limited. Type FPL cables are suitable for general use. General use does not include use in plenums, ducts, risers, or other environmental air-handling spaces unless the cables are installed in metallic raceways as permitted in Section 300.22. Type FPL may also have additional letters to designate special uses. The letter P stands for "plenum," and the letter R stands for "riser." Cables with 2-hour fire ratings intended to enhance survivability in a fire may also be marked with CI for circuit integrity. (See Figure 4.13 and the related discussion under the heading "Wiring Methods and Materials for NPLFA Circuits.")

Using the marking scheme described, we see that Type FPLP can be used in ducts, plenums, or other space for environmental air without enclosing the cables in a raceway. However, Section 760.3(B) refers to Section 300.22 generally, and Section 300.22(A) says no such wiring is permitted in ducts that handle loose stock or vapor, with or without raceways. Therefore, Section 760.61(A) has the effect of modifying Section 300.22 only with respect to the types of wiring that will be permitted, but does not allow FPLP to be installed in a duct that handles loose stock or vapor.

The P suffix indicates that the cable has been tested and passes a plenum flame test that measures flame propagation and smoke density. To be suitable for a plenum application, such cables should not spread a fire rapidly or produce a lot of smoke that would be transported by the air-handling system. This marking makes selection of cables easy for installers and identification of cables to verify compliance easy for inspectors.

Cables marked FPLR are intended for riser use. The R suffix means that the cable passes a riser flame test. To be installed in a riser, a cable should be slow in propagating flames so that a fire will not spread from floor to floor by following the cable. Riser cables are required where cables are installed in vertical runs in shafts or penetrating more than one floor, unless the cables are installed in metal raceways or in fireproof shafts with firestops at each floor. Riser applications in one- and two-family dwellings do not require riser cables.

Two special types of cables also use the preceding markings: coaxial cables and insulated continuous line-type fire detectors. Line-type detectors take various forms, but the electrical cable type responds to heat by a change in resistance that can be sensed by a control circuit. An example of an application for continuous line-type fire detectors is shown in Figure 4.18, which illustrates the line-type detector being used in a cable tray.

Section 760.61(D) and Table 760.61 provide the permitted substitutions for FPL cable types. These substitutions are illustrated in Figure 760.61, which is reproduced here as Figure 4.19.

An issue relating to PLFA circuits that has sometimes been controversial is the application of conduit fill limits to these circuits. The primary reference that requires compliance with the conduit fill limits of Table 1 is in Section 300.17. Section 760.3 excludes Article 760 circuits from the requirements of Article 300 except where there is a specific reference to some section of Article 300. For NPLFA circuits, Section 300.17 is specifically referenced in Section 760.25. So for NPLFA circuits and PLFA circuits using NPLFA wiring methods, conduit fill clearly applies. Section 300.17 is referenced again in Sections 760.28(A) and 760.28(B) for NPLFA circuits mixed with other circuits. However, Section 300.17 is not mentioned in Part I or Part III of Article 760, so there is no reference to apply Section 300.17 to PLFA circuit cables. For this reason, many people argue that conduit fill limits do not apply to PLFA circuits using PLFA cables.

However, suppose we need to run some FPL cable in a raceway in a riser application. If we choose to use electrical metallic tubing (EMT), for example, we must follow the requirements for the instal-

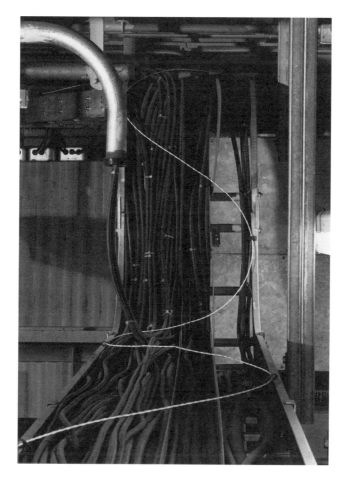

FIGURE 4.18 Typical line-type continuous fire detector. (Source: *National Fire Alarm Code® Handbook,* NFPA, 1999, Exhibit 2.8; courtesy of Protectowire Co., Hanover, MA)

lation of the EMT found in Article 358. Section 358.22 says cables are required to meet the requirements of Table 1 in Chapter 9. A similar rule is found in every other raceway article. Because Article 760 exempted PLFA circuits from the rules of Article 300 but not from the rest of the rules in Chapter 3 of the *Code,* it appears that conduit fill limits do apply to PLFA cables.

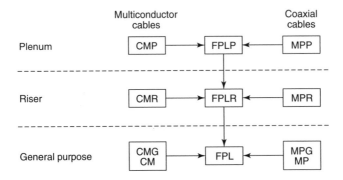

Type CM—Communications wires and cables
Type FPL—Power-limited fire alarm cables
Type MP—Multipurpose cables (coaxial cables only)

A→B Cable A shall be permitted to be used in place of cable B.

FIGURE 4.19 Cable substitution hierarchy. (Source: *National Electrical Code®*, NFPA, 2002, Figure 760.61)

PLFA Circuit Separations

As for NPLFA circuits, PLFA circuits are permitted to use special installation rules and wiring methods, but they are also subject to special restrictions. The primary restriction is the requirement for circuit separations. These restrictions are found in Section 760.55. The types of circuits that can be mixed without separations and the restrictions on these circuits are found in Section 760.56.

Special treatment is accorded PLFA circuits because they are supplied from power-limited sources. These sources are carefully specified, designed, and tested to be sure that the energy limits can be relied on. In order to be certain that the energy limits will not be compromised, energy-limited circuits must be separated from higher powered circuits. This separation provides reasonable assurance that the power-limited circuits will remain power-limited circuits.

The basic separation requirement that PLFA circuits will not be in close proximity to or in contact with higher energy circuits is found in Section 760.55(A). However, in subsections 760.55(B) through 760.55(J), Section 760.55 provides a range of ways to maintain the separations. Power-limited fire alarm conductors may be separated by a barrier or a raceway within an enclosure and still occupy the same overall enclosure. Other specific methods of maintaining separations are provided for specific types of installations. Generally, a physical separation of at least 2 inches is required unless separation is provided by metal raceways, metal cable sheaths, Type UF cables, or continuous fixed nonconductors. In hoistways the separation is usually provided by raceways. Special rules are provided in Article 620 for fire alarm conductors that must be in traveling cables in order to remain connected to an elevator car. We do not discuss all the specific possibilities in this book. The reader can find a method of separation for most situations in Section 760.55.

Maintaining separations is a special problem in enclosures that PLFA conductors must enter to connect to the same equipment as higher powered circuits. This situation arises often in fire alarm control panels and similar locations, where, for example, a PLFA control circuit must connect to the same controller as the power circuit it controls. In such cases, two options are provided in Section 760.55(D). First, physical separations may be provided within the enclosure, but the requirement is reduced to ¼ inch from the usual 2 inches. This requirement that the conductors be "routed to maintain a minimum of 6 mm (0.25 in.) . . ." is shown in Figure 4.20. In order to maintain this separation, the conductors must usually be fastened in place or otherwise contained in a wire management system of some sort.

The second option is available where the circuit conductors operate at no more than 150 volts to ground. In this case, two more options are available: The conductors can be installed as NPLFA circuit conductors or the jacket of an FPL or a permitted substitute cable may be used for separation (see Figure 4.21). NPLFA conductors are permitted to occupy the same enclosure with higher power conductors if they connect to the same equipment. Where PLFA or substitute cable jackets are used for separation, the conductors that are not contained in the cable jacket still must meet the minimum ¼ inch spacing requirement.

When PLFA conductors must enter an enclosure with power conductors and the enclosure has only one opening, the limited energy circuits must be separated from the higher powered circuits by

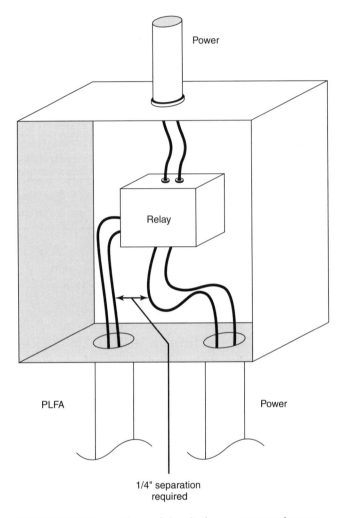

Power

Relay

PLFA   Power

1/4" separation
required

FIGURE 4.20 Separations of circuits in common enclosures.

a nonconductor, such as flexible tubing, that is firmly fixed in place. This requirement is found in Section 760.55(E).

Section 760.56 lists the circuit types that may be mixed in an enclosure and the conditions under which they may be mixed. Essentially, circuits of the same class may be in the same cable, or in the same raceway or other enclosure. For the purposes of this rule, Class 3 circuits and communications circuits are considered to be of es-

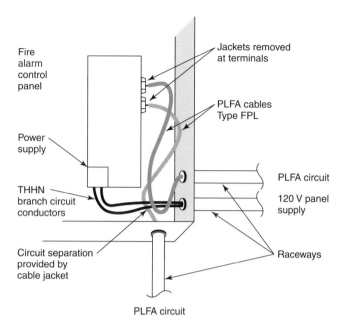

Fire alarm control panel

Jackets removed at terminals

PLFA cables Type FPL

Power supply

PLFA circuit

120 V panel supply

THHN branch circuit conductors

Circuit separation provided by cable jacket

Raceways

PLFA circuit

FIGURE 4.21 Separations provided by cable jackets.

sentially the same class as PLFA circuits. They need not be functionally associated.

Class 2 and PLFA circuits may be in the same cable, enclosure, or raceway if all the circuits are insulated to PLFA requirements. Typically, this condition would mean using PLFA, Class 3, or acceptable substitute cables for both the Class 2 and the PLFA circuit conductors. Class 2 cable is only required to have 150 volt insulation, but all other limited energy cables are required to be insulated for 300 volts.

PLFA cables may also be mixed in the same raceway or enclosure, but not the same cable, with low-power network-powered broadband communications circuits.

RELATED REQUIREMENTS

At the beginning of this chapter we discussed the relationship between the *NEC* and other codes with regard to the requirements for fire alarm systems. The *NEC* and Article 760 are primarily installation standards, so Article 760 does not say where a fire alarm system is required, what elements are required, or where they are to be located. Once those requirements have been established, Article 760 provides requirements for wiring of the system. Indirectly, the *NEC*, as an installation standard, also incorporates any installation or wiring instructions included in the listing or labeling of the fire alarm panel, the initiating devices, and the notification appliances. However, as noted in Section 760.1, Fine Print Note, it is *NFPA 72, National Fire Alarm Code,* that covers the "monitoring for integrity requirements of fire alarm systems." The requirements of *NFPA 72* are, for the most part, beyond the scope of this book. Nevertheless, certain issues in monitoring for integrity are directly related to wiring practices, and readers should be aware of some of these requirements. Therefore, the objective of this section is to examine some of the issues related to wiring and monitoring for integrity.

Section 90.1 says the purpose of the *NEC* is to provide safe installations. Safety in this context is primarily safety from the hazards to persons and property arising from the use of electricity. Convenience, proper function, good service, and adequacy are not primary concerns of the *NEC*. Despite this disclaimer, the *NEC* does require proper function of certain systems such as overcurrent devices, grounding systems, and emergency power supplies. When the failure of a system results in a hazard to life or property, the *NEC* steps in to cover those types of failures. However, in the case of fire alarm systems, the rules that are intended to be sure a fire alarm system is in working order and able to warn occupants of a fire in a building are not specifically addressed in the *NEC*. Many of these requirements are covered instead by *NFPA 72*.

Monitoring for Integrity

NFPA 72 addresses many issues in the design and construction of fire alarm systems and equipment. For example, Section 1-5.2.3 requires fire alarm systems to have at least two independent and reliable sources of power. Often these sources are a normal branch cir-

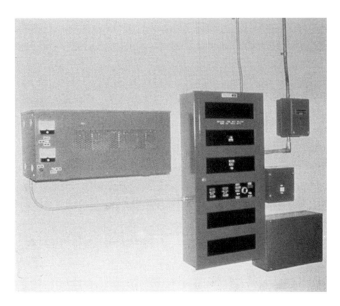

FIGURE 4.22   Two power sources required by *NFPA 72*. (Source: *National Fire Alarm Code® Handbook*, NFPA, 1999, Exhibit 1.36)

cuit and an integral battery (as shown in Figure 4.22), or they may be the two sources required for most emergency power systems. Figure 4.22 shows a fire alarm panel supplied by a normal branch circuit on the right, with a separated battery cabinet on the left. Batteries are frequently mounted inside a panel, but the load will determine how much battery capacity is required.

Monitoring for integrity is intended to provide a sort of electrical or electronic supervision of a fire alarm system, its power supplies, wiring, initating devices, and notification appliances so that the system can be relied on to serve its function when the need arises. The means of monitoring may vary from one system to another, and is integral to the control panel, but for the monitoring to work and be reliable, the wiring must be installed properly, and the connections to devices and appliances must be made correctly.

According to Section 1-5.8.1 of *NFPA 72*,

> All means of interconnecting equipment, devices, appliances, and wiring connections shall be monitored for the integrity of the interconnecting conductors or equivalent path so that the occurrence

of a single open or a single ground-fault condition in the installation conductors or other signaling channels and their restoration to normal shall be automatically indicated within 200 seconds.

The basic way that this requirement is met is by imposing a small current on a fire alarm circuit so that normal conditions are indicated by the continuous flow of the current. The current value is determined by a device at the end of the circuit such as a resistor as shown in Figure 4.23. The resistor is often called an EOL (end-of-line) resistor, although it may be located in the control panel. As illustrated in Figure 4.23, an open conductor or the disconnection of a device will cause the current to stop, and other faults will cause the current to change in other ways depending on the design of the circuit and the nature of the fault. Such changes typically cause a trouble signal at the control panel. While this arrangement has been the traditional way of monitoring indicating device circuits and notifying appliance circuits, there are other methods.

One common method of monitoring used in many modern systems involves multiplexing and the use of what is called a *signaling*

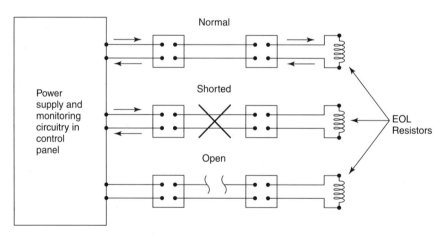

Top: Normal current is determined by EOL resistor
Middle: High current due to shorted conductors
(may be interpreted as an alarm)
Bottom: No current due to open circuit

FIGURE 4.23 Function of an end-of-line (EOL) resistor.

line circuit. In one version of this method, multiple signals are sent on the same circuit. Each device has its own address or name. The control panel sends a series of signals, one to each device by name, *interrogating* the device. Each device sends a return signal (or *response*) that indicates its status. The control panel can then locate defective or missing devices or defects in wiring. Using this method, the control panel will also know, when it receives an alarm signal, which device sent the signal. In many older systems, like the system described in the preceding paragraph, the panel could only locate trouble or alarms by zone or circuit. Both old and new general schemes have advantages, but in order for either to work, the installations and connections to devices must be made correctly.

One problem for some installers, especially those who have done a lot of power wiring, is realizing that the connections to devices must usually be made in a particular fashion. If the wires to a device are pigtailed or T-tapped, the loss or failure of a device may not be sensed because the wiring is designed to go into and out of each device. Figure 4.24 illustrates the practice of T-tapping. If either of the tap wires to the separate device were to be broken, the panel would be unable to detect the damage because the circuit through the EOL device remains intact. For the same reason, wires connecting to a

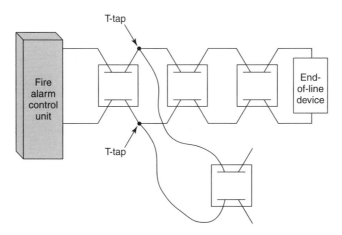

FIGURE 4.24 Illustration of T-**tapping. (Source:** *National Fire Alarm Code®* *Handbook,* **NFPA, 1999, Exhibit 3.1)**

device must be cut so that the continuity of the circuit is dependent on the device, as shown in Figure 4.25. On the one hand, splicing a conductor on a device is often considered a bad practice, or even a prohibited practice for a grounded conductor in some power circuits, but is necessary to the monitoring function of many fire alarm circuits. On the other hand, in certain types of signaling line circuits, T-tapping as shown in Figure 4.24 is acceptable because the interrogation routine identifies each device. Therefore a lost connection to the separate device will be identified because the device will not individually respond to the panel interrogation. Signaling line circuits do not rely entirely on a continuous current flow to prove the integrity of the circuit. The specific circuit design and wiring instructions must be followed explicitly for each installation and may vary somewhat from one installation to another.

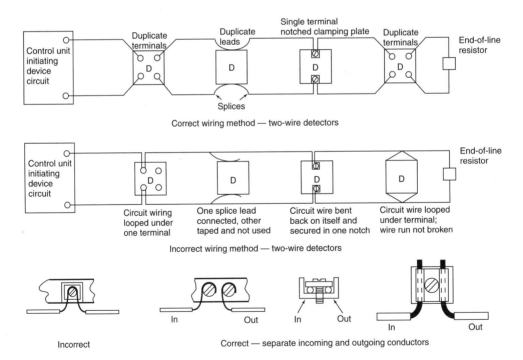

FIGURE 4.25 Fire alarm circuit wiring connections. [Source: *National Fire Alarm Code®* *Handbook*, NFPA, 1999, Figure A.2.1.3.4(a)]

Circuit Designs

Another factor that directly affects the system integrity and performance is the class and style of the circuit. Initiating device circuits, notification appliance circuits, and signaling line circuits are permitted to be one of two classes: Class A or Class B. The class of a circuit is determined by or describes the way a circuit will perform during a single fault of a specific type. How the circuit performs depends on the control panel and the way the wiring is laid out and connected. Class A and Class B circuits are defined or specified in Section 3-4.2.1 of *NFPA 72*. Essentially, a Class A circuit is a circuit that is capable of transmitting an alarm signal during a single open or a nonsimultaneous single ground fault on a circuit conductor. A Class B circuit is incapable of transmitting an alarm beyond the location of the same sort of single fault. Basic examples of Class A and Class B circuits are shown in Figure 4.26. In the Class B circuit, the panel can detect the break shown, but the devices beyond the break can no longer transmit an alarm because they are effectively disconnected from the panel. The Class A circuit maintains contact with every device even when a single break occurs. In the event of a single open or single ground fault on a circuit, the panel must indicate the fault as a trouble condition as required by Section 1-5.8.1, as quoted in the previous section herein.

The definitions of Class A and Class B circuits recognize that a single fault on a Class B circuit will disable all devices or appliances downstream from the fault. However, in order to minimize the possibility that damage to a cable or a raceway will defeat the intended Class A circuit operation, the outgoing and return conductors are required to be routed separately. (Many jurisdictions refer to wiring routed this way as a "McCulloh loop," especially where the conductors are installed in conduit. However, McCulloh systems refer to something completely different in *NFPA 72*. See Section 5-5.3.3 for more information.) Of course, where the conductors enter the panel, they will necessarily be in close proximity, and exceptions are provided to Section 3-4.2.2.2 to deal with this and similar issues. Appendix A of *NFPA 72* provides recommendations for minimum separation distances of 1 foot in vertical runs and 4 feet in

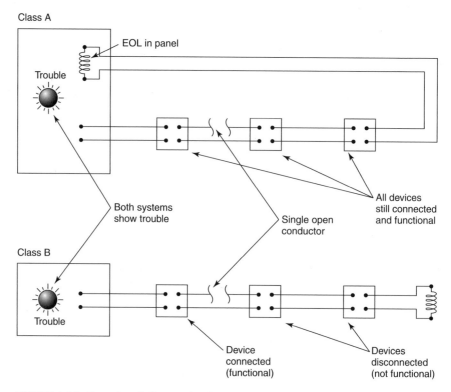

Class A

EOL in panel

Trouble

Both systems
show trouble

Single open
conductor

All devices
still connected
and functional

Class B

Trouble

Device
connected
(functional)

Devices
disconnected
(not functional)

FIGURE 4.26 Class A and Class B circuit styles.

horizontal runs, but local jurisdictions or differences in building construction or installation of wiring methods may permit or require lesser or greater separations. For example, open cables run horizontally in a dropped ceiling are more susceptible to damage than conduits run vertically and embedded in a concrete wall.

In addition to class, circuits are also grouped by style. Initiating device circuits can be Style A, B, C, D, or E. Notification appliance circuits can be Style W, X, Y, or Z. Signaling line circuits can be Style 0.5, 1, 2, 3, 3.5, 4, 4.5, 5, 6, or 7. These styles provide a description of the ability of a circuit to meet performance requirements during specified fault conditions. The performance requirements for the various styles are specified in Tables 3-5, 3-6, and 3-7 of *NFPA 72*. Figure 4.27 shows Table 3-5 reproduced as an example. Notice that in Table 3-5, a wire-to-wire short may produce either a trouble signal

Class	B			B			B			A			A		
Style	A			B			C			D			Eα		
Abnormal condition	Alarm	Trouble	Alarm receipt capability during abnormal condition	Alarm	Trouble	Alarm receipt capability during abnormal condition	Alarm	Trouble	Alarm receipt capability during abnormal condition	Alarm	Trouble	Alarm receipt capability during abnormal condition	Alarm	Trouble	Alarm receipt capability during abnormal condition
	1	2	3	4	5	6	7	8	9	10	11	12	13	14	15
Single open	—	X	—	—	X	—	—	X	—	—	X	X	—	X	X
Single ground	—	X	—	—	X	R	—	X	R	—	X	R	—	X	R
Wire-to-wire short	X	—	—	X	—	—	—	X	—	X	—	—	—	X	—
Loss of carrier (if used)/ channel interface	—	—	—	—	—	—	—	X	—	—	—	—	—	X	—

R = Required capacity
X = Indication required at protected premises and as required by Chapter 5
α = Style exceeds minimum requirements of Class A

FIGURE 4.27 Sample style table for Class A and Class B circuits. (Source: *National Fire Alarm Code® Handbook,* NFPA, 1999, Table 3.5)

or an alarm signal, but in no style is the panel expected to receive an alarm signal from a circuit with a wire-to-wire short. The style of a circuit is mostly dependant on the control panel, but in some cases the wiring arrangement may be affected. In many installations, notification appliance circuits may be installed as one class and initiating device circuits as another class.

Again, the point of this whole discussion of related requirements is that fire alarm circuits vary in design and installation requirements. Installation and wiring issues are not all covered by the *NEC*. Circuit classes, styles, and performance characteristics are described and defined in *NFPA 72*. The method of monitoring for integrity and the choice of a class and style of a circuit are not specified by *NFPA 72*, but are selected by an owner or designer, or, perhaps, specified by an enforcement authority or insurance company. To be certain that a fire alarm system will work as required, the specific wiring and connection diagrams must be carefully followed in each case.

SUMMARY

Article 760 covers power-limited (PLFA) and non-power-limited (NPLFA) fire alarm circuits. These two circuit classifications covered by Article 760 are defined by the voltage and current limitations of their power sources, which are usually a part of a fire alarm control panel. Article 760 provides special rules that modify the general requirements of the first four chapters of the *NEC*.

This chapter covers the various ways that the general rules are modified for PLFA and NPLFA circuits. Both types of circuits have special rules with regard to wire sizes, derating factors, overcurrent protection, and separation from other circuits, but NPLFA circuits generally use the same insulation types and wiring methods permitted for ordinary circuits, although some special cable types and their uses are specified in Article 760. Different insulation types and wiring methods are permitted for PLFA circuits. Finally, this chapter also discusses some related wiring requirements found in *NFPA 72, National Fire Alarm Code*.

Fire Alarm Systems

I. General

760.1 Scope. This article covers the installation of wiring and equipment of fire alarm systems including all circuits controlled and powered by the fire alarm system.

> FPN No. 1: Fire alarm systems include fire detection and alarm notification, guard's tour, sprinkler waterflow, and sprinkler supervisory systems. Circuits controlled and powered by the fire alarm system include circuits for the control of building systems safety functions, elevator capture, elevator shutdown, door release, smoke doors and damper control, fire doors and damper control and fan shutdown, but only where these circuits are powered by and controlled by the fire alarm system. For further information on the installation and monitoring for integrity requirements for fire alarm systems, refer to the *NFPA 72*-1999, *National Fire Alarm Code®*.

> FPN No. 2: Class 1, 2, and 3 circuits are defined in Article 725.

760.2 Definitions. For purposes of this article, the following definitions apply.

Abandoned Fire Alarm Cable. Installed fire alarm cable that is not terminated at equipment other than a connector and not identified for future use with a tag.

Fire Alarm Circuit. The portion of the wiring system between the load side of the overcurrent device or the power-limited supply and the connected equipment of all circuits powered and controlled by the fire alarm system. Fire alarm circuits are classified as either non–power-limited or power-limited.

Fire Alarm Circuit Integrity (CI) Cable. Cable used in fire alarm systems to ensure continued operation of critical circuits during a specified time under fire conditions.

Non–Power-Limited Fire Alarm Circuit (NPLFA). A fire alarm circuit powered by a source that complies with 760.21 and 760.23.

Source: NFPA 70, *National Electrical Code®*, NFPA, Quincy, MA, 2002 edition.

Power-Limited Fire Alarm Circuit (PLFA). A fire alarm circuit powered by a source that complies with 760.41.

760.3 Locations and Other Articles. Circuits and equipment shall comply with 760.3(A) through (F). Only those sections of Article 300 referenced in this article shall apply to fire alarm systems.

(A) Spread of Fire or Products of Combustion. Section 300.21. The accessible portion of abandoned fire alarm cables shall not be permitted to remain.

(B) Ducts, Plenums, and Other Air-Handling Spaces. Section 300.22, where installed in ducts or plenums or other spaces used for environmental air.

Exception: As permitted in 760.30(B)(1) and (2) and 760.61(A).

(C) Hazardous (Classified) Locations. Articles 500 through 516 and Article 517, Part IV, where installed in hazardous (classified) locations.

(D) Corrosive, Damp, or Wet Locations. Sections 110.11, 300.6, and 310.9 where installed in corrosive, damp, or wet locations.

(E) Building Control Circuits. Article 725 where building control circuits (e.g., elevator capture, fan shutdown) are associated with the fire alarm system.

(F) Optical Fiber Cables. Where optical fiber cables are utilized for fire alarm circuits, the cables shall be installed in accordance with Article 770.

760.5 Access to Electrical Equipment Behind Panels Designed to Allow Access. Access to electrical equipment shall not be denied by an accumulation of conductors and cables that prevents removal of panels, including suspended ceiling panels.

760.6 Mechanical Execution of Work. Fire alarm circuits shall be installed in a neat and workmanlike manner. Cables and conductors installed exposed on the surface of ceiling and sidewalls shall be supported by structural components of the building in such a manner that the cable or conductors will not be damaged by normal building use. Such cables shall be attached to structural components by straps, staples, hangers, or similar fittings designed and installed so as not to damage the cable. The installation shall also conform with 300.4(D).

760.7 Fire Alarm Circuits Extending Beyond One Building. Power-limited fire alarm circuits that extend beyond one building and run outdoors either shall meet the installation requirements of Parts II, III, and IV of Article 800 or shall meet the installation requirements of Part I of Article 300. Non–power-limited fire alarm circuits that extend beyond one building and run outdoors shall meet the installation requirements of Part I of Article 300 and the applicable sections of Part I of Article 225.

760.9 Fire Alarm Circuit and Equipment Grounding. Fire alarm circuits and equipment shall be grounded in accordance with Article 250.

760.10 Fire Alarm Circuit Identification. Fire alarm circuits shall be identified at terminal and junction locations, in a manner that will prevent unintentional interference with the signaling circuit during testing and servicing.

760.15 Fire Alarm Circuit Requirements. Fire alarm circuits shall comply with the following parts of this article.

(A) Non–Power-Limited Fire Alarm (NPLFA) Circuits. See Parts I and II.

(B) Power-Limited Fire Alarm (PLFA) Circuits. See Parts I and III.

II. Non–Power-Limited Fire Alarm (NPLFA) Circuits

760.21 NPLFA Circuit Power Source Requirements. The power source of non–power-limited fire alarm circuits shall comply with Chapters 1 through 4, and the output voltage shall not be more than 600 volts, nominal. These circuits shall not be supplied through ground-fault circuit interrupters.

> FPN: See 210.8(A)(5), Exception No. 3 for receptacles in dwelling-unit unfinished basements that supply power for fire alarm systems.

760.23 NPLFA Circuit Overcurrent Protection. Overcurrent protection for conductors 14 AWG and larger shall be provided in accordance with the conductor ampacity without applying the derating factors of 310.15 to the ampacity calculation. Overcurrent protection shall not exceed 7 amperes for 18 AWG conductors and 10 amperes for 16 AWG conductors.

Exception: Where other articles of this Code permit or require other overcurrent protection.

760.24 NPLFA Circuit Overcurrent Device Location. Overcurrent devices shall be located at the point where the conductor to be protected receives its supply.

Exception No. 1: Where the overcurrent device protecting the larger conductor also protects the smaller conductor.

Exception No. 2: Transformer secondary conductors. Non–power-limited fire alarm circuit conductors supplied by the secondary of a single-phase transformer that has only a 2-wire (single-voltage) secondary shall be permitted to be protected by overcurrent protection provided by the primary (supply) side of the transformer, provided the protection is in accordance with 450.3 and does not exceed the value determined by multiplying the secondary conductor ampacity by the secondary-to-primary transformer voltage ratio. Transformer secondary conductors other than 2-wire shall not be considered to be protected by the primary overcurrent protection.

Exception No. 3: Electronic power source output conductors. Non–power-limited circuit conductors supplied by the output of a single-phase, listed electronic power source, other than a transformer, having only a 2-wire (single-voltage) output for connection to non–power-limited circuits shall be permitted to be protected by overcurrent protection provided on the input side of the electronic power source, provided this protection does not exceed the value determined by multiplying the non–power-limited circuit conductor ampacity by the output-to-input voltage ratio. Electronic power source outputs, other than 2-wire (single voltage), connected to non–power-limited circuits shall not be considered to be protected by overcurrent protection on the input of the electronic power source.

> FPN: A single-phase, listed electronic power supply whose output supplies a 2-wire (single-voltage) circuit is an example of a non–power-limited power source that meets the requirements of 760.21.

760.25 NPLFA Circuit Wiring Methods. Installation of non–power-limited fire alarm circuits shall be in accordance with 110.3(B), 300.11(A), 300.15, 300.17, and other appropriate articles of Chapter 3.

Exception No. 1: As provided in 760.26 through 760.30.

Exception No. 2: Where other articles of this Code require other methods.

760.26 Conductors of Different Circuits in Same Cable, Enclosure, or Raceway.

(A) Class 1 with NPLFA Circuits. Class 1 and non–power-limited fire alarm circuits shall be permitted to occupy the same cable, enclosure, or raceway without regard to whether the individual circuits are alternating current or direct current, provided all conductors are insulated for the maximum voltage of any conductor in the enclosure or raceway.

(B) Fire Alarm with Power-Supply Circuits. Power-supply and fire alarm circuit conductors shall be permitted in the same cable, enclosure, or raceway only where connected to the same equipment.

760.27 NPLFA Circuit Conductors.

(A) Sizes and Use. Only copper conductors shall be permitted to be used for fire alarm systems. Size 18 AWG and 16 AWG conductors shall be permitted to be used, provided they supply loads that do not exceed the ampacities given in Table 402.5 and are installed in a raceway, an approved enclosure, or a listed cable. Conductors larger than 16 AWG shall not supply loads greater than the ampacities given in 310.15, as applicable.

(B) Insulation. Insulation on conductors shall be suitable for 600 volts. Conductors larger than 16 AWG shall comply with Article 310. Conductors 18 AWG and 16 AWG shall be Type KF-2, KFF-2, PAFF, PTFF, PF, PFF, PGF, PGFF, RFH-2, RFHH-2, RFHH-3, SF-2, SFF-2, TF, TFF, TFN, TFFN, ZF, or ZFF. Conductors with other types and thickness of insulation shall be permitted if listed for non–power-limited fire alarm circuit use.

FPN: For application provisions, see Table 402.3.

(C) Conductor Materials. Conductors shall be solid or stranded copper.

Exception to (B) and (C): Wire Types PAF and PTF shall be permitted only for high-temperature applications between 90°C (194°F) and 250°C (482°F).

760.28 Number of Conductors in Cable Trays and Raceways, and Derating.

(A) NPLFA Circuits and Class 1 Circuits. Where only non–power-limited fire alarm circuit and Class 1 circuit conductors are in a raceway, the number of conductors shall be determined in accordance with 300.17. The derating factors given in 310.15(B)(2)(a) shall apply if such conductors carry continuous load in excess of 10 percent of the ampacity of each conductor.

(B) Power-Supply Conductors and Fire Alarm Circuit Conductors. Where power-supply conductors and fire alarm circuit conductors are permitted in a raceway in accordance with 760.26, the number of conductors shall be determined in accordance with 300.17. The derating factors given in 310.15(B)(2)(a) shall apply as follows:

(1) To all conductors where the fire alarm circuit conductors carry continuous loads in excess of 10 percent of the ampacity of each conductor and where the total number of conductors is more than three
(2) To the power-supply conductors only, where the fire alarm circuit conductors do not carry continuous loads in excess of 10 percent of the ampacity of each conductor and where the number of power-supply conductors is more than three

(C) Cable Trays. Where fire alarm circuit conductors are installed in cable trays, they shall comply with 392.9 through 392.11.

760.30 Multiconductor NPLFA Cables. Multiconductor non–power-limited fire alarm cables that meet the requirements of 760.31 shall be permitted to be used on fire alarm circuits operating at 150 volts or less and shall be installed in accordance with 760.30(A) and (B).

(A) NPLFA Wiring Method. Multiconductor non–power-limited fire alarm circuit cables shall be installed in accordance with 760.30(A)(1), (A)(2), and (A)(3).

(1) Exposed or Fished in Concealed Spaces. In raceway or exposed on surface of ceiling and sidewalls or fished in concealed spaces. Cable splices or terminations shall be made in listed fittings, boxes, enclosures, fire alarm devices, or utilization equipment. Where installed exposed, cables shall be adequately supported and installed in such a way that maximum protection against physical damage is afforded by building construction such as baseboards, door frames, ledges, and so forth. Where located within 2.1 m (7 ft) of the floor, cables shall be securely fastened in an approved manner at intervals of not more than 450 mm (18 in.).

(2) Passing Through a Floor or Wall. In metal raceway or rigid nonmetallic conduit where passing through a floor or wall to a height of 2.1 m (7 ft) above the floor unless adequate protection can be afforded by building construction such as detailed in 760.30(A)(1) or unless an equivalent solid guard is provided.

(3) In Hoistways. In rigid metal conduit, rigid nonmetallic conduit, intermediate metal conduit, liquidtight flexible nonmetallic tubing, or electrical metallic tubing where installed in hoistways.

Exception: As provided for in 620.21 for elevators and similar equipment.

(B) Applications of Listed NPLFA Cables. The use of non–power-limited fire alarm circuit cables shall comply with 760.30(B)(1) through (B)(4).

(1) Ducts and Plenums. Multiconductor non–power-limited fire alarm circuit cables, Types NPLFP, NPLFR, and NPLF, shall not be installed exposed in ducts or plenums.

FPN: See 300.22(B).

(2) Other Spaces Used for Environmental Air. Cables installed in other spaces used for environmental air shall be Type NPLFP.

Exception No. 1: Types NPLFR and NPLF cables installed in compliance with 300.22(C).

Exception No. 2: Other wiring methods in accordance with 300.22(C) and conductors in compliance with 760.27(C).

(3) Riser. Cables installed in vertical runs and penetrating more than one floor or cables installed in vertical runs in a shaft shall be Type NPLFR. Floor penetrations requiring Type NPLFR shall contain only cables suitable for riser or plenum use.

Exception No. 1: Type NPLF or other cables that are specified in Chapter 3 and are in compliance with 760.27(C) and encased in metal raceway.

Exception No. 2: Type NPLF cables located in a fireproof shaft having firestops at each floor.

FPN: See 300.21 for firestop requirements for floor penetrations.

(4) Other Wiring Within Buildings. Cables installed in building locations other than the locations covered in 760.30(B)(1), (B)(2), and (B)(3) shall be Type NPLF.

Exception No. 1: Chapter 3 wiring methods with conductors in compliance with 760.27(C).

Exception No. 2: Type NPLFP or Type NPLFR cables shall be permitted.

760.31 Listing and Marking of NPLFA Cables. Non–power-limited fire alarm cables installed as wiring within buildings shall be listed in accordance with 760.31(A) and (B) and as being resistant to the spread of fire in accordance with 760.31(C) through (F), and shall be marked in accordance with 760.31(G).

(A) NPLFA Conductor Materials. Conductors shall be 18 AWG or larger solid or stranded copper.

(B) Insulated Conductors. Insulated conductors shall be suitable for 600 volts. Insulated conductors 14 AWG and larger shall be one of the types listed in Table 310.13 or one that is identified for this use. Insulated conductors 18 AWG and 16 AWG shall be in accordance with 760.27.

(C) Type NPLFP. Type NPLFP non–power-limited fire alarm cable for use in other space used for environmental air shall be listed as being suitable for use in other space used for environmental air as described in 300.22(C) and shall also be listed as having adequate fire-resistant and low smoke-producing characteristics.

FPN: One method of defining low smoke-producing cable is by establishing an acceptable value of the smoke produced when tested in accordance with NFPA 262-1999, *Standard Method of Test for Flame Travel and Smoke of Wires and Cables for Use in Air-Handling Spaces*, to a maximum peak optical density of 0.5 and a maximum average optical density of 0.15. Similarly, one method of defining fire-resistant cables is by establishing a maximum allowable flame travel distance of 1.52 m (5 ft) when tested in accordance with the same test.

(D) Type NPLFR. Type NPLFR non–power-limited fire alarm riser cable shall be listed as being suitable for use in a vertical run in a shaft or from floor to floor and shall also be listed as having fire-resistant characteristics capable of preventing the carrying of fire from floor to floor.

> FPN: One method of defining fire-resistant characteristics capable of preventing the carrying of fire from floor to floor is that the cables pass ANSI/UL 1666-1997, *Test for Flame Propagation Height of Electrical and Optical-Fiber Cables Installed Vertically in Shafts.*

(E) Type NPLF. Type NPLF non–power-limited fire alarm cable shall be listed as being suitable for general-purpose fire alarm use, with the exception of risers, ducts, plenums, and other space used for environmental air, and shall also be listed as being resistant to the spread of fire.

> FPN No. 1: One method of defining *resistant to the spread of fire* is that the cables do not spread fire to the top of the tray in the vertical-tray flame test in ANSI/UL 1581-1991, *Reference Standard for Electrical Wires, Cables and Flexible Cords.*

> FPN No. 2: Another method of defining *resistant to the spread of fire* is for the damage (char length) not to exceed 1.5 m (4 ft 11 in.) when performing the CSA vertical flame test for cables in cable trays, as described in CSA C22.2 No. 0.3-M-1985, *Test Methods for Electrical Wires and Cables.*

(F) Fire Alarm Circuit Integrity (CI) Cable. Cables suitable for use in fire alarm systems to ensure survivability of critical circuits during a specified time under fire conditions shall be listed as circuit integrity (CI) cable. Cables identified in 760.31(C), (D), and (E) that meet the requirements for circuit integrity shall have the additional classification using the suffix "CI" (for example, NPLFP-CI, NPLFR-CI, and NPLF-CI).

> FPN No. 1: This cable may be used for fire alarm circuits to comply with the survivability requirements of *NFPA 72-1999, National Fire Alarm Code®*, 3-4.2.2.2, 3-8.4.1.1.4, and 3-8.4.1.3.3.3(3), that the cable maintain its electrical function during fire conditions for a defined period of time.

> FPN No. 2: One method of defining circuit integrity (CI) cable is by establishing a minimum 2-hour fire resistance rating for the cable when tested in accordance with UL 2196-1995, *Standard for Tests of Fire Resistive Cables.*

(G) NPLFA Cable Markings. Multiconductor non–power-limited fire alarm cables shall be marked in accordance with Table 760.31(G). Non–power-limited fire alarm circuit cables shall be permitted to be marked with a maximum usage voltage rating of 150 volts. Cables that are listed for circuit integrity shall be identified with the suffix "CI" as defined in 760.31(F).

Table 760.31(G) NPLFA Cable Markings

Cable Marking	Type	Reference
NPLFP	Non–power-limited fire alarm circuit cable for use in other space used for environmental air	760.31(C) and (G)
NPLFR	Non–power-limited fire alarm circuit riser cable	760.31(D) and (G)
NPLF	Non–power-limited fire alarm circuit cable	760.31(E) and (G)

Note: Cables identified in 760.31(C), (D), and (E) and meeting the requirements for circuit integrity shall have the additional classification using the suffix "CI" (for example, NPLFP-CI, NPLFR-CI, and NPLF-CI).

FPN: Cable types are listed in descending order of fire resistance rating.

III. Power-Limited Fire Alarm (PLFA) Circuits

760.41 Power Sources for PLFA Circuits. The power source for a power-limited fire alarm circuit shall be as specified in 760.41(A), (B), or (C). These circuits shall not be supplied through ground-fault circuit interrupters.

FPN No. 1: Table 12(A) and 12(B) in Chapter 9 provide the listing requirements for power-limited fire alarm circuit sources.

FPN No. 2: See 210.8(A)(5), Exception No. 3, for receptacles in dwelling-unit unfinished basements that supply power for fire alarm systems.

(A) Transformers. A listed PLFA or Class 3 transformer.

(B) Power Supplies. A listed PLFA or Class 3 power supply.

(C) Listed Equipment. Listed equipment marked to identify the PLFA power source.

FPN: Examples of listed equipment are a fire alarm control panel with integral power source; a circuit card listed for use as a PLFA source, where used as part of a listed assembly; a current-limiting impedance, listed for the purpose or part of a listed product, used in conjunction with a non–power-limited transformer or a stored energy source, for example, storage battery, to limit the output current.

760.42 Circuit Marking. The equipment shall be durably marked where plainly visible to indicate each circuit that is a power-limited fire alarm circuit.

> FPN: See 760.52(A), Exception No. 3 where a power-limited circuit is to be reclassified as a non–power-limited circuit.

760.51 Wiring Methods on Supply Side of the PLFA Power Source. Conductors and equipment on the supply side of the power source shall be installed in accordance with the appropriate requirements of Part II and Chapters 1 through 4. Transformers or other devices supplied from power-supply conductors shall be protected by an overcurrent device rated not over 20 amperes.

Exception: The input leads of a transformer or other power source supplying power-limited fire alarm circuits shall be permitted to be smaller than 14 AWG, but not smaller than 18 AWG, if they are not over 300 mm (12 in.) long and if they have insulation that complies with 760.27(B).

760.52 Wiring Methods and Materials on Load Side of the PLFA Power Source. Fire alarm circuits on the load side of the power source shall be permitted to be installed using wiring methods and materials in accordance with either 760.52(A) or (B).

(A) NPLFA Wiring Methods and Materials. Installation shall be in accordance with 760.25, and conductors shall be solid or stranded copper.

Exception No. 1: The derating factors given in 310.15(B)(2)(a) shall not apply.

Exception No. 2: Conductors and multiconductor cables described in and installed in accordance with 760.27 and 760.30 shall be permitted.

Exception No. 3: Power-limited circuits shall be permitted to be reclassified and installed as non–power-limited circuits if the power-limited fire alarm circuit markings required by 760.42 are eliminated and the entire circuit is installed using the wiring methods and materials in accordance with Part II, Non–Power-Limited Fire Alarm Circuits.

> FPN: Power-limited circuits reclassified and installed as non–power-limited circuits are no longer power-limited circuits, regardless of the continued connection to a power-limited source.

(B) PLFA Wiring Methods and Materials. Power-limited fire alarm conductors and cables described in 760.71 shall be installed as detailed in 760.52(B)(1), (2), or (3) of this section. Devices shall be installed in accordance with 110.3(B), 300.11(A), and 300.15.

(1) Exposed or Fished in Concealed Spaces. In raceway or exposed on the surface of ceiling and sidewalls or fished in concealed spaces. Cable splices or terminations shall be made in listed fittings, boxes, enclosures, fire alarm devices, or utilization equipment. Where installed exposed, cables shall be adequately supported and installed in such a way that maximum protection against physical damage is afforded by building construction such as baseboards, door frames, ledges, and so forth. Where located within 2.1 m (7 ft) of the floor, cables shall be securely fastened in an approved manner at intervals of not more than 450 mm (18 in.).

(2) Passing Through a Floor or Wall. In metal raceways or rigid nonmetallic conduit where passing through a floor or wall to a height of 2.1 m (7 ft) above the floor, unless adequate protection can be afforded by building construction such as detailed in 760.52(B)(1) or unless an equivalent solid guard is provided.

(3) In Hoistways. In rigid metal conduit, rigid nonmetallic conduit, intermediate metal conduit, or electrical metallic tubing where installed in hoistways.

Exception No. 1: As provided for in 620.21 for elevators and similar equipment.

Exception No. 2: Other wiring methods and materials installed in accordance with the requirements of 760.3 shall be permitted to extend or replace the conductors and cables described in 760.71 and permitted by 760.52(B).

760.54 Installation of Conductors and Equipment in Cables, Compartments, Cable Trays, Enclosures, Manholes, Outlet Boxes, Device Boxes, and Raceways for Power-Limited Circuits. Conductors and equipment for power-limited fire alarm circuits shall be installed in accordance with 760.55 through 760.58.

760.55 Separation from Electric Light, Power, Class 1, NPLFA, and Medium Power Network-Powered Broadband Communications Circuit Conductors.

(A) General. Power-limited fire alarm circuit cables and conductors shall not be placed in any cable, cable tray, compartment, enclosure, manhole, outlet box, device box, raceway, or similar fitting with conductors of electric light, power, Class 1, non–power-limited fire alarm circuits, and medium power network-powered broadband communications circuits unless permitted by 760.55(B) through (G).

(B) Separated by Barriers. Power-limited fire alarm circuit cables shall be permitted to be installed together with Class 1, non–power-limited fire alarm, and medium power network-powered broadband communications circuits where they are separated by a barrier.

(C) Raceways Within Enclosures. In enclosures, power-limited fire alarm circuits shall be permitted to be installed in a raceway within the enclosure to separate them from Class 1, non–power-limited fire alarm, and medium power network-powered broadband communications circuits.

(D) Associated Systems Within Enclosures. Power-limited fire alarm conductors in compartments, enclosures, device boxes, outlet boxes, or similar fittings shall be permitted to be installed with electric light, power, Class 1, non–power-limited fire alarm, and medium power network-powered broadband communications circuits where they are introduced solely to connect the equipment connected to power-limited fire alarm circuits, and comply with either of the following conditions:

(1) The electric light, power, Class 1, non–power-limited fire alarm, and medium power network-powered broadband communications circuit conductors are routed to maintain a minimum of 6 mm (0.25 in.) separation from the conductors and cables of power-limited fire alarm circuits.

(2) The circuit conductors operate at 150 volts or less to ground and also comply with one of the following:

   a. The fire alarm power-limited circuits are installed using Type FPL, FPLR, FPLP, or permitted substitute cables, provided these power-limited cable conductors extending beyond the jacket are separated by a minimum of 6 mm (0.25 in.) or by a nonconductive sleeve or nonconductive barrier from all other conductors.

   b. The power-limited fire alarm circuit conductors are installed as non–power-limited circuits in accordance with 760.25.

(E) Enclosures with Single Opening. Power-limited fire alarm circuit conductors entering compartments, enclosures, device boxes, outlet boxes, or similar fittings shall be permitted to be installed with electric light, power, Class 1 non–power-limited fire alarm, and medium power network-powered broadband communications circuits where they are introduced solely to connect the equipment connected to power-limited fire alarm circuits or to other circuits controlled by the fire alarm system to which the other conductors in the enclosure are connected. Where power-limited fire alarm circuit conductors must enter an enclosure that is provided with a single opening, they shall be permitted to enter through a single fitting (such as a tee), provided the conductors are separated from the conductors of the other circuits by a continuous and firmly fixed nonconductor, such as flexible tubing.

(F) In Hoistways. In hoistways, power-limited fire alarm circuit conductors shall be installed in rigid metal conduit, rigid nonmetallic conduit, intermediate metal conduit, liquidtight flexible nonmetallic conduit, or electrical metallic tubing. For elevators or similar equipment, these conductors shall be permitted to be installed as provided in 620.21.

(G) Other Applications. For other applications, power-limited fire alarm circuit conductors shall be separated by at least 50 mm (2 in.) from conductors of any electric light, power, Class 1, non–power-limited fire alarm, or medium power network-powered broadband communications circuits unless one of the following conditions is met:

(1) Either (a) all of the electric light, power, Class 1, non–power-limited fire alarm, and medium power network-powered broadband communications circuit conductors or (b) all of the power-limited fire alarm circuit conductors are in a raceway or in metal-sheathed, metal-clad, nonmetallic-sheathed, or Type UF cables.
(2) All of the electric light, power, Class 1 non–power-limited fire alarm, and medium power network-powered broadband communications circuit conductors are permanently separated from all of the power-limited fire alarm circuit conductors by a continuous and firmly fixed nonconductor, such as porcelain tubes or flexible tubing, in addition to the insulation on the conductors.

760.56 Installation of Conductors of Different PLFA Circuits, Class 2, Class 3, and Communications Circuits in the Same Cable, Enclosure, or Raceway.

(A) Two or More PLFA Circuits. Cable and conductors of two or more power-limited fire alarm circuits, communications circuits, or Class 3 circuits shall be permitted within the same cable, enclosure, or raceway.

(B) Class 2 Circuits with PLFA Circuits. Conductors of one or more Class 2 circuits shall be permitted within the same cable, enclosure, or raceway with conductors of power-limited fire alarm circuits, provided that the insulation of the Class 2 circuit conductors in the cable, enclosure, or raceway is at least that required by the power-limited fire alarm circuits.

(C) Low-Power Network-Powered Broadband Communications Cables and PLFA Cables. Low-power network-powered broadband communications circuits shall be permitted in the same enclosure or raceway with PLFA cables.

760.57 Support of Conductors. Power-limited fire alarm circuit conductors shall not be strapped, taped, or attached by any means to the exterior of any conduit or other raceway as a means of support.

760.58 Conductor Size. Conductors of 26 AWG shall be permitted only where spliced with a connector listed as suitable for 26 AWG to 24 AWG or larger conductors that are terminated on equipment or where the 26 AWG conductors are terminated on equipment listed as suitable for 26 AWG conductors. Single conductors shall not be smaller than 18 AWG.

760.59 Current-Carrying Continuous Line-Type Fire Detectors.

(A) Application. Listed continuous line-type fire detectors, including insulated copper tubing of pneumatically operated detectors, employed for both detection and carrying signaling currents shall be permitted to be used in power-limited circuits.

(B) Installation. Continuous line-type fire detectors shall be installed in accordance with 760.42 through 760.52 and 760.54.

760.61 Applications of Listed PLFA Cables. PLFA cables shall comply with the requirements described in either 760.61(A), (B), or (C) or where cable substitutions are made as shown in 760.61(D).

(A) Plenum. Cables installed in ducts, plenums, and other spaces used for environmental air shall be Type FPLP. Abandoned cables shall not be permitted to remain. Types FPLP, FPLR, and FPL cables installed in compliance with 300.22 shall be permitted.

(B) Riser. Cables installed in risers shall be as described in either (1), (2), or (3):

(1) Cables installed in vertical runs and penetrating more than one floor, or cables installed in vertical runs in a shaft, shall be Type FPLR. Floor penetrations requiring Type FPLR shall contain only cables suitable for riser or plenum use. Abandoned cables shall not be permitted to remain.
(2) Other cables shall be installed in metal raceways or located in a fireproof shaft having firestops at each floor.
(3) Type FPL cable shall be permitted in one- and two-family dwellings.

FPN: See 300.21 for firestop requirements for floor penetrations.

(C) Other Wiring Within Buildings. Cables installed in building locations other than those covered in 760.61(A) or (B) shall be as described in either (1), (2), (3), or (4).

(1) Type FPL shall be permitted.
(2) Cables shall be permitted to be installed in raceways.
(3) Cables specified in Chapter 3 and meeting the requirements of 760.71(A) and (B) shall be permitted to be installed in nonconcealed spaces where the exposed length of cable does not exceed 3 m (10 ft).
(4) A portable fire alarm system provided to protect a stage or set when not in use shall be permitted to use wiring methods in accordance with 530.12.

(D) Fire Alarm Cable Uses and Permitted Substitutions. The uses and permitted substitutions for fire alarm cables listed in Table 760.61 shall be considered suitable for the purpose and shall be permitted.

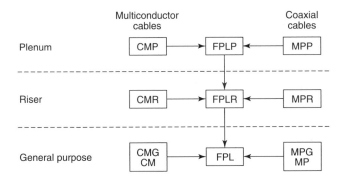

Type CM—Communications wires and cables
Type FPL—Power-limited fire alarm cables
Type MP—Multipurpose cables (coaxial cables only)

A→B Cable A shall be permitted to be used in place of cable B.

26 AW6 minimum

Figure 760.61 Cable substitution hierarchy.

FPN: For information on multipurpose cables (Types MPP, MPR, MPG, MP) and communications cables (Types CMP, CMR, CMG, CM), see 800.51.

760.71 Listing and Marking of PLFA Cables and Insulated Continuous Line-Type Fire Detectors. Type FPL cables installed as wiring within buildings shall be listed as being resistant to the spread of fire and other criteria in accordance with 760.71(A) through (H) and shall be marked in accordance with 760.71(I). Insulated continuous line-type fire detectors shall be listed in accordance with 760.71(J).

Table 760.61 Cable Uses and Permitted Substitutions

			Permitted Substitutions	
Cable Type	Use	References	Multiconductor	Coaxial
FPLP	Power-limited fire alarm plenum cable	760.61(A)	CMP	MPP
FPLR	Power-limited fire alarm riser cable	760.61(B)	CMP, FPLP, CMR	MPP, MPR
FPL	Power-limited fire alarm cable	760.61(C)	CMP, FPLP, CMR, FPLR, CMG, CM	MPP, MPR, MPG, MP

(A) Conductor Materials. Conductors shall be solid or stranded copper.

(B) Conductor Size. The size of conductors in a multiconductor cable shall not be smaller than 26 AWG. Single conductors shall not be smaller than 18 AWG.

(C) Ratings. The cable shall have a voltage rating of not less than 300 volts.

(D) Type FPLP. Type FPLP power-limited fire alarm plenum cable shall be listed as being suitable for use in ducts, plenums, and other space used for environmental air and shall also be listed as having adequate fire-resistant and low smoke-producing characteristics.

> FPN: One method of defining low smoke-producing cable is by establishing an acceptable value of the smoke produced when tested in accordance with NFPA 262-1999, *Standard Method of Test for Flame Travel and Smoke of Wires and Cables for Use in Air-Handling Spaces*, to a maximum peak optical density of 0.5 and a maximum average optical density of 0.15. Similarly, one method of defining fire-resistant cables is by establishing maximum allowable flame travel distance of 1.52 m (5 ft) when tested in accordance with the same test.

(E) Type FPLR. Type FPLR power-limited fire alarm riser cable shall be listed as being suitable for use in a vertical run in a shaft or from floor to floor and shall also be listed as having fire-resistant characteristics capable of preventing the carrying of fire from floor to floor.

> FPN: One method of defining fire-resistant characteristics capable of preventing the carrying of fire from floor to floor is that the cables pass the requirements of ANSI/UL 1666-1997, *Standard Test for Flame Propagation Height of Electrical and Optical-Fiber Cable Installed Vertically in Shafts*.

(F) Type FPL. Type FPL power-limited fire alarm cable shall be listed as being suitable for general-purpose fire alarm use, with the exception of risers, ducts, plenums, and other spaces used for environmental air and shall also be listed as being resistant to the spread of fire.

> FPN: One method of defining *resistant to the spread of fire* is that the cables do not spread fire to the top of the tray in the vertical-tray flame test in ANSI/UL 1581-1991, *Reference Standard for Electrical Wires, Cables and Flexible Cords*. Another method of defining *resistant to the spread of fire* is for the damage (char length) not to exceed 1.5 m (4 ft 11 in.) when performing the CSA vertical flame test—cables in cable trays, as described in CSA C22.2 No. 0.3-M-1985, *Test Methods for Electrical Wires and Cables*.

(G) Fire Alarm Circuit Integrity (CI) Cable. Cables suitable for use in fire alarm systems to ensure survivability of critical circuits during a specified time under fire conditions shall be listed as circuit integrity (CI) cable. Cables identified in 760.71(D), (E), and (F) that meet the requirements for circuit integrity shall have the additional classification using the suffix "CI" (for example, FPLP-CI, FPLR-CI, and FPL-CI).

FPN No. 1: This cable is used for fire alarm circuits as one method of complying with the survivability requirements of *NFPA 72-1999, National Fire Alarm Code*, 3-4.2.2.2, 3-8.4.1.1.4, and 3-8.4.1.3.3.3(3), that the cable maintain its electrical function during fire conditions for a defined period of time.

FPN No. 2: One method of defining circuit integrity (CI) cable is by establishing a minimum 2-hour fire resistance rating for the cable when tested in accordance with UL 2196-1995, *Standard for Tests of Fire Resistive Cables*.

(H) Coaxial Cables. Coaxial cables shall be permitted to use 30 percent conductivity copper-covered steel center conductor wire and shall be listed as Type FPLP, FPLR, or FPL cable.

(I) Cable Marking. The cable shall be marked in accordance with Table 760.71(I). The voltage rating shall not be marked on the cable. Cables that are listed for circuit integrity shall be identified with the suffix CI as defined in 760.71(G).

FPN: Voltage ratings on cables may be misinterpreted to suggest that the cables may be suitable for Class 1, electric light, and power applications.

Exception: Voltage markings shall be permitted where the cable has multiple listings and voltage marking is required for one or more of the listings.

Table 760.71(I) Cable Markings

Cable Marking	Type	Listing References
FPLP	Power-limited fire alarm plenum cable	760.71(D) and (I)
FPLR	Power-limited fire alarm riser cable	760.71(E) and (I)
FPL	Power-limited fire alarm cable	760.71(F) and (I)

Note: Cables identified in (D), (E), and (F) meeting the requirements for circuit integrity shall have the additional classification using the suffix "CI" (for example, FPLP-CI, FPLR-CI, and FPL-CI).

FPN: Cable types are listed in descending order of fire-resistance rating.

(J) Insulated Continuous Line-Type Fire Detectors. Insulated continuous line-type fire detectors shall be rated in accordance with 760.71(C), listed as being resistant to the spread of fire in accordance with 760.71(D) through (F), marked in accordance with 760.71(I), and the jacket compound shall have a high degree of abrasion resistance.

CHAPTER

Optical Fiber Cables

Article 770 first appeared in the *National Electrical Code®* in the 1984 *NEC®*. At that time, optical fiber cables were also added to the scope of the *Code* itself in Article 90. The intent was to cover those installations of fiber-optic cables that included electrical conductors or that were installed along with electrical conductors. It was recognized that optical fiber cables were increasingly being used in conjunction with electrical conductors or replacing electrical conductors in applications such as signaling and control circuits. These uses were in addition to the communications installations that were the primary early applications of fiber optics.

When the topic first appeared in the *Code,* some people thought optical fiber cables belonged in Chapter 8 because the primary uses had been or were thought to be in communications. In fact, early applications of fiber optics in industrial controls, such as for so-called data highways, required the services of communications installers because they were the only ones with experience in terminating, splicing, and testing optical fiber cables. However, applications and connection and splicing methods have changed dramatically over the years, and many more installers have been trained in optical fiber installations. Fiber optics are used in many applications other than communications, such as industrial controls, signaling, and computer interconnections for data transfer.

177

Although the scope of Article 770 was originally limited to optical fiber cables that were installed along with electrical conductors, the scope of the *NEC* included optical fiber cables generally. In 1990, Article 770 was changed to match the scope of the *NEC*, and since then Article 770 has covered optical fibers with or without electrical conductors.

Optical fiber technologies do pose certain hazards to installers. Probably the most obvious hazard is the laser light used as a light source. The *NEC* does not cover this hazard. Optical fibers may be included in communications or other limited energy cables such as Class 2, Class 3, and power-limited fire alarm cables, but under the listing standards for these cables, the optical fibers are limited to carrying only optical energy that has been ruled not hazardous to the human body. The limits of optical energy are set by the Food and Drug Administration of the U.S. Department of Health. The limits are the Class I laser-radiation levels found in the *Code of Federal Regulations*, 21 *CFR* Part 1040. So this hazard is addressed by a regulatory body, but not by the *NEC*.

A primary reason for including a seemingly nonelectrical technology under the provisions of the *NEC* is based on the fact that the optical fiber cables assume many of the functions primarily fulfilled by metallic conductors in the past. Furthermore, many optical fiber installations are actually mixed installations in which a control, data, or communications signal is converted to and from electrical energy. The interface where these conversions take place, along with the related cabling on each side of the interface, are most efficiently and safely handled by the same designers and installers. Some installations use cables that include both optical fibers and electrical conductors in the same cable. Also, many or most optical fiber cables are installed in raceways that are the same or nearly the same as the raceways used for electrical conductors.

Although optical fibers are not wiring as such, many of the wiring methods from the *NEC* are used in installations of optical fiber cables. Where the optical fibers are intermixed in a cable, raceway, or other enclosure with electrical conductors, the rules for the electrical conductors must be followed. In other cases, optical fibers that are not associated with electrical conductors are essentially the same as foreign systems, such as pneumatic tubing, and should be isolated from the electrical installation. To consider another case, where conductors of limited energy circuits are mixed with optical fibers, there is little risk from the limited energy circuits energizing even a conductive element in the optical fiber cable, and no electrical risk associated with the optical fiber. Therefore, the *NEC* does differentiate between different types of optical fiber cables and does provide separation requirements for different types of cables and different types of electrical circuits.

Article 770 modifies certain general requirements of the *NEC* to deal with the unique aspects of optical fiber cables. In some ways Article 770 is similar to Article 725, which is covered in Chapter 3 of this book, but in other ways it is more similar to Article 800, which is covered in Chapter 6 herein. In the following sections of the present chapter, we examine the ways optical fiber cables are categorized, how the general rules for wiring methods are modified and how they are similar for optical fiber cables, and how optical fiber cables should be selected and applied to specific installations in buildings.

DEFINITIONS

Since Article 770 is included in the first seven chapters of the *NEC*, the definitions in Article 100 apply to Article 770. However, Article 770 does include some definitions of terms that have different meanings with regard to optical fiber installations than they do generally, under the definitions of Article 100 or as used elsewhere in the *NEC*. Two terms that mean something different in Article 770 are *exposed* and *point of entrance*.

In Article 100 and as used generally in the *NEC*, "exposed" may apply to live parts or to wiring methods. Exposed live parts are energized parts that can be inadvertently touched because they are not "suitably guarded, isolated, or insulated." Exposed wiring methods are "on or attached to the surface or behind panels designed to allow access." "Exposed" may also be used in other contexts to mean visible or subject to damage. However, in Article 770, "exposed" takes on the same meaning that it has in Chapter 8 of the *Code:* "The circuit is in such a position that, in case of failure of supports or insulation, contact with another circuit may result."

"Point of entrance" is not defined in Article 100, but the term is used in Articles 225, 230, and 240 to describe the point were conductors enter a building or other structure by passing through a wall or other defining boundary. Under Article 770, "point of entrance" takes on the meaning it has in communications installations: "The point at which the wire or cable emerges from an external wall, from a concrete floor slab, or from a rigid metal conduit or an intermediate metal conduit grounded to an electrode in accordance with 800.40(B)." Elsewhere in the *NEC,* except in communications, the point of entrance is at the boundary, but in Article 770 and in communications, the point of entry can be extended into a building by placing the entering cable in a properly grounded rigid metal conduit (RMC) or intermediate metal conduit (IMC) raceway. In that case, the point of entrance becomes the point where the cable emerges from the raceway, as shown in Figure 5.1.

APPLICATION OF OTHER CODE REQUIREMENTS

As in other articles covering limited energy circuits, Article 770 excludes itself from the provisions of Article 300 except where a section of Article 300 is referenced within Article 770. For example, in Sections 770.3(A) and 770.3(B), the requirements of Sections 300.21 and 300.22 are specifically referenced. These sections require installations of optical fiber cables to meet certain requirements that apply to other wiring methods, specifically, to not increase the spread of fire or smoke and to comply with the restrictions on wiring in ducts, plenums, and other air-handling spaces. Since these requirements are really about physical characteristics of the materials as combustibles and how they are installed, rather than about electrical hazards as such, the requirements should apply to any materials, electrical or not.

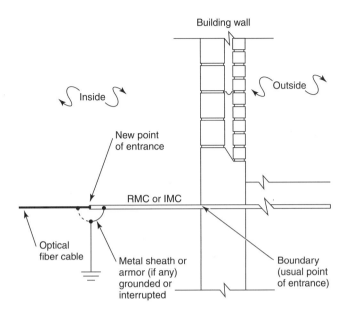

FIGURE 5.1 Point of entrance.

CLASSIFICATIONS OF CABLES

Optical fiber cables are grouped into three types based on their construction: nonconductive, conductive, or composite. Cables may have strength members to increase the tensile strength and reduce pulling stresses on the optical fibers. Strength members may be metallic (conductive) or nonconductive. Cables may have shields or armor, which typically are metallic. Cables may also include electrical conductors along with the optical fibers. The three types recognize these differences and the requirements in other sections of Article 770 are tailored to the differences.

Section 770.5 provides descriptions of the three cable types. Nonconductive cables contain no metallic members and no other conductive elements such as shields. Conductive cables contain metallic strength members or other metallic elements such as vapor barriers, armor, or sheaths, but no current-carrying conductors. Composite cables contain current-carrying electrical conductors along with optical fibers, and perhaps other metallic elements such as strength members or shields.

Nonconductive and conductive optical fiber cables are listed and identified as OFN (optical fiber nonconductive) or OFC (optical fiber conductive) types. A nonconductive cable is shown in Figure 5.2. Composite cables are listed according to the type of electrical conductors they contain, with the additional marking "-OF." For example, a cable intended for use with Class 2 circuits in a riser application that also contains one or more optical fibers would be marked "CL2R-OF." It may also have other markings to indicate additional listings or other intended uses such as, perhaps, "direct burial." A composite cable listed as Type MC is shown in Figure 5.3.

WIRING METHODS

Certain requirements that apply to wiring generally also apply to installations of optical fiber cables. Sections 770.7 and 770.8 contain rules that also apply to all limited energy circuits and are especially applicable with regard to cables that are run without raceways. As

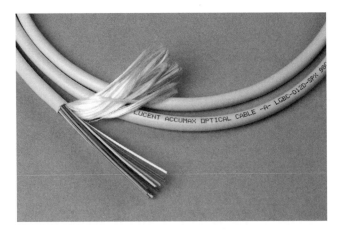

FIGURE 5.2 Nonconductive optical fiber cable. (Source: *National Electrical Code®* *Handbook*, NFPA, 2002, Exhibit 770.1; courtesy of AT&T Network Cable Systems)

FIGURE 5.3 Composite optical fiber cable, Type MC. (Source: *National Electrical Code® Handbook*, NFPA, 2002, Exhibit 770.2; courtesy of AFC Cable Systems)

required by Section 770.7, cables are not allowed to accumulate to the point where they prevent the removal of panels such as acoustical panels in suspended grid ceilings, and unused abandoned cables must be removed. Section 770.8 requires that optical fiber cables exposed on wall or ceiling surfaces be supported from the structure, be installed in a "neat and workmanlike manner," and be protected from damage by screws or nails when run parallel to framing members. Figure 5.4 illustrates the misuse of a ceiling grid as both a grid for cable support and with an accumulation of cables that interferes with access to the equipment in the space above the removable panels.

Wiring methods for optical fibers include cables alone or cables in raceways. There are some cables that can be substituted for other cables, such as a plenum cable that can be substituted for riser or general purpose cable, but the only real substitute for an optical fiber cable is another optical fiber cable. Because composite cables are listed according their construction and the types of electrical conductors they include, so the substitutions for a particular cable

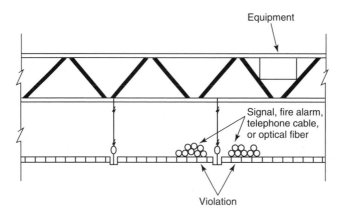

FIGURE 5.4 Misuse of ceiling grid.

type, such as CL3 for CL2 or CM (communications) for CL3 can be followed when electrical conductors are in the same cable with optical fibers.

Section 770.6 covers raceways generally and allows the use of *NEC* Chapter 3 wiring methods according to the Chapter 3 rules for that wiring method. The exception is listed optical fiber raceways. These raceways do not have their own article. Sections 770.51(E), (F), and (G) describe the three types of these raceways: plenum, riser, and general purpose. The permitted uses for plenum raceways are the same as the permitted uses of plenum cables. Riser and general-purpose raceways are treated the same as riser and general-purpose cables respectively. However, as a raceway, certain installation requirements apply to the raceways that do not necessarily apply to cable.

Optical fiber raceways are installed under the rules for ENT (electrical nonmetallic tubing) covered by Article 362. Only Sections 362.24 through 362.56 are actually referenced in Section 770.6, Exception. Section 362.22, which is not referenced, covers conduit fill. This lack of referencing does not necessarily mean that conduit fill requirements do not apply. References within each of the individual raceway articles do require compliance with the conduit fill requirements in Table 1, Chapter 9 of the *NEC*, but even these rules are modified for optical fiber cables. Section 770.6 says where conduit fill limitations apply and where they do not. Conduit fill limits apply when nonconductive optical fiber cables are installed in a raceway along with current-carrying conductors. Conduit fill limits do not apply when optical fiber cables are installed without any current-carrying conductors in the same raceway. Article 770 does not say how to treat conductive optical fiber cables installed with current-carrying conductors. However, conductive cables cannot be mixed in a raceway with non-power-limited circuits, and a rule about how to make a prohibited installation should not be expected. Nevertheless, conductive optical fiber cables may be installed in the same raceway with power-limited conductors. Therefore, the applicability of conduit fill requirements to such installations can be determined by consulting the rules for the specific power-limited conductors. For example, if the current-carrying conductors are Class 2 or Class 3 circuits, conduit fill limits must be observed because the limits apply to the Class 2 and Class 3 conductors, according to Section 725.3(A). However, conduit fill limitations do not apply to communications cables, according to Section 800.48, Exception. So where conductive optical fiber cables are installed with current-carrying communications circuits, conduit fill limits do not apply. These examples are illustrated in Figure 5.5. Note that this discussion disregards the controversy about whether or not communications conductors are "current carrying."

There are other practical considerations for optical fiber cables. Optical fiber cables can be easily damaged by excessive pulling tensions. Conductive or nonconductive strength members are included in some cables partly to keep pulling tensions from being transmitted to the optical fibers. Pulling tensions are increased when raceways are overfilled. Therefore, fill may need to be kept below the maximum fill permitted by Table 1 in Chapter 9 of the *NEC*, to avoid damage to the cables during installation. Most manufacturers will provide maximum pulling tensions for their cables. Many manufacturers also provide information on how to calculate or predict pulling tension in a given installation.

If cables are to be used, with or without raceways, they must comply with the listing and marking requirements of Section 770.50.

Cable Types:

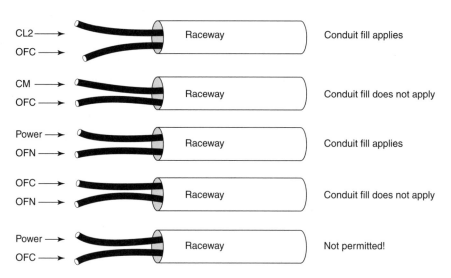

	Raceway	
CL2 → OFC →		Conduit fill applies
CM → OFC →		Conduit fill does not apply
Power → OFN →		Conduit fill applies
OFC → OFN →		Conduit fill does not apply
Power → OFC →		Not permitted!

FIGURE 5.5 Conduit fill requirements for optical fiber cables.

This section requires optical fiber cables to be listed and marked if they are to be used inside of buildings. Three exceptions allow unlisted cables to (1) enter a building and run up to 50 feet and terminate in an enclosure, (2) enter a building from outside and remain inside an RMC or IMC conduit that is grounded to an electrode, and (3) enter a building and stay in a raceway. Exception No. 1 applies to all optical fiber cables, Exception No. 2 applies only to conductive cables, and Exception No. 3 applies only to nonconductive cables. As given in the definition of "point of entrance," Exception No. 2 actually extends the point of entrance to the end of the grounded RMC or IMC. Exception No. 1 starts at the point of entrance. Thus both Exception No. 1 and Exception No. 2 can be used in the same installation as illustrated by Figure 5.6.

The exceptions to Section 770.50 are particularly useful with entrance cables because many cables intended for use only outside are not listed, but they must run some distance into a building so they can be terminated.

Conductive cables are marked OFC and nonconductive cables are marked OFN. Either of these markings may have an additional letter *P* for plenum, or *R* for riser. Thus, OFCP and OFNP cables are listed for use in ducts, plenums, and other space for environmental air, and OFCR and OFNR cables are listed for use in riser applications. The letter *G* denotes that OFCG and OFNG cables are listed for general-purpose use; OFC and OFN cables are also for general-purpose use.

The actual locations where each of these cable types and their corresponding raceway types may be used are given in Section 770.53. As may be expected, the locations where various cable types may be used are based only on the fire-resistance and smoke-production properties of the cables because the optical energy does not present a risk of either shock or fire. We find in Sections 770.53(D) and (E) that any of the listed cable types can be used in hazardous locations or cable trays. However, if the cable tray is in a so-called plenum ceiling, the cables would have to be suitable

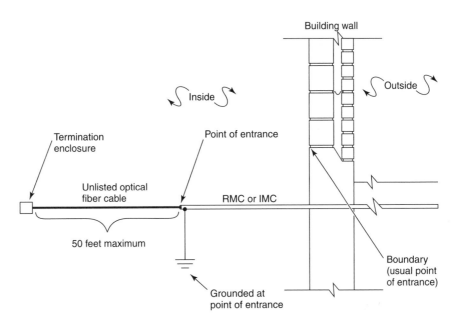

FIGURE 5.6 Use of unlisted cables from point of entrance.

for other spaces for environmental air. Sections 770.53(A) and (B) reiterate the requirement for removing abandoned cables that also appears in Section 770.3(A).

Cables installed in ducts, plenums, or other space for environmental air must be suitable for that use or be installed in appropriate raceways, including plenum optical fiber raceways. Other cable types may also be used in air-handling spaces if they are enclosed in raceways, except that plenum optical fiber raceways may only contain OFCP or OFNP cables as shown in Figure 5.7.

Cables installed in riser applications may be either riser or plenum rated. Where riser optical fiber raceways are used, only riser or plenum cables may be used in the raceway. When metal raceways or fireproof shafts are used for risers, any of the cable types may be used. General-purpose cable types may be used in riser applications without raceways in one- and two-family dwellings.

Cable types intended for general-purpose applications—Types OFC, OFN, OFCG, and OFNG—can be used anywhere in buildings except where plenum or riser cables are required. These cable types may be installed without raceways, or with raceways, including listed general-purpose optical fiber raceways.

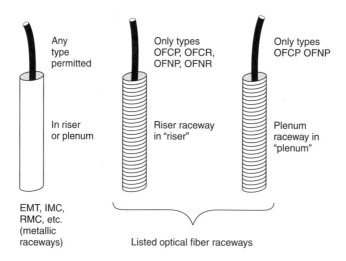

FIGURE 5.7 Use of listed optical fiber raceways.

SEPARATIONS

At first glance, the separation requirements for optical fiber cables appear to be the same or similar to the separation requirements for limited energy circuits. Separations and permitted mixing of circuits depend on the type of cable and the types of circuits as well as where the cables and circuits may be mixed. Three types of cables have been established in Article 770. By definition, the only optical fiber cables that could have electrical circuits in the same cable are composite cables. The other two types are conductive and nonconductive. Two types of electrical circuits are considered, non-power-limited circuits up to 600 volts and power-limited circuits. The exceptions to Section 770.52 actually create a third possibility— circuits exceeding 600 volts. Two places where conductors may be combined with optical fibers are considered: (1) raceways and cable trays where cables are run together but do not terminate and (2) cabinets, outlet boxes, panels, and similar enclosures where electrical conductors terminate.

Composite cables may contain optical fibers along with power, lighting, Class 1, and similar non-power-limited circuit conductors only where they are functionally associated and the electrical circuits are limited to 600 volts. Functional association is required if the optical fibers and electrical conductors are to occupy the same cable. Composite cables containing Class 1 and similar non-power-limited conductors may also occupy the same raceway or termination enclosure with other circuits of the same types, and functional association is not required. Section 770.52, Exception No. 3 allows composite cables to include conductors of circuits over 600 volts in industrial occupancies where only qualified persons will service the installation. Optical fibers may also share a composite cable with power-limited circuit conductors. Section 770.52 does not say anything about mixing the power-limited conductors with non-power-limited conductors in a cable, raceway, or enclosure. Those restrictions are found in the articles applicable to the type of power-limited circuit. Article 770 provides only the restrictions for optical fiber cables.

Nonconductive cables do not contain any electrical circuit conductors by definition. Nonconductive cables may occupy the same raceway or cable tray with non-power-limited circuits of up to 600 volts, but may *not* occupy the same cabinet, panel, or other termination enclosure with non-power-limited circuit conductors. Section 770.52, Exceptions No. 1 and No. 2 permit nonconductive cables to occupy a termination enclosure only if the fibers and conductors are functionally associated or if they are in factory- or field-assembled control panels. Exception No. 3 permits a nonconductive cable to be with circuits over 600 volts in industrial occupancies where only qualified persons will service the installation. Nonconductive cables may be mixed in any raceway or enclosure with power-limited circuits and no functional association is required.

Conductive optical fiber cables may not occupy the same raceway, cable tray, or other enclosure with any conductors except power-limited circuit conductors.

To summarize these separation requirements:

- Composite cables are the least restricted. However, functional association is required for non-power-limited circuits that occupy the same cable raceway, cable, or enclosure with optical fibers, and the separation requirements for the circuit types still apply. For example, a composite cable could contain Class 1 and power conductors along with optical fibers, but they would all have to be functionally associated to comply with both Articles 770 and 725. A composite cable generally could not contain optical fibers, Class 2, and power conductors because the separation requirements for the Class 2 cables would be violated.

- Nonconductive cables are more restricted than composite cables in that they may occupy a common raceway or cable tray, but generally may not occupy a common termination enclosure unless they are functionally associated with the non-power-limited conductors.

- Conductive cables are the most restricted because they are not permitted to occupy any raceways or enclosures with non-power-limited conductors.

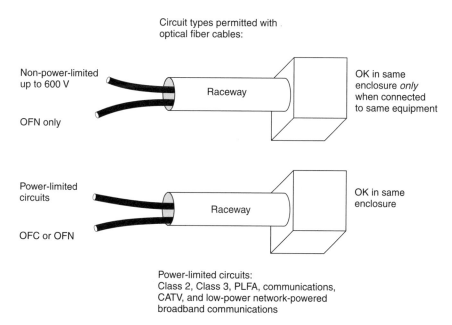

FIGURE 5.8 Optical fibers and power-limited circuits.

- As shown in Figure 5.8, no restrictions are placed on any of the optical fiber cables being mixed with power-limited circuits other than the restrictions that apply to the power-limited circuits themselves.

GROUNDING

Section 770.53(C) requires the non-current-carrying conductive members of optical fiber cables to be grounded in accordance with Article 250. This requirement could only be applied to composite or conductive cables because nonconductive cables do not have conductive members. Composite cables are covered in Article 250 based on the type of circuit that is enclosed. Composite cables may contain power-limited or non-power-limited circuits. The non-current-carrying parts associated with most non-power-limited circuits require grounding, but similar parts associated with many power-limited circuits do not require grounding. These requirements are covered in detail in Chapter 3 of this book.

Article 250 does not generally require non-current-carrying parts of nonelectrical equipment to be grounded. A few such parts are required to be grounded as covered in Section 250.116. However, Section 250.116 does not list applications likely to include uses for optical fibers. If conductive optical fiber cable is associated with some electrical equipment where it could become energized, such as in the same raceway with a power-limited circuit, then grounding of the conductive elements may be required. However, since conductive cables are generally required to be isolated from circuits that are required to be grounded, the conductive element may not require grounding at all. Because most of the grounding requirements and methods are based on circuit ratings (overcurrent device ratings), some circuits that could energize the conductive element would have to be identified to get instructions from Article 250 on how to ground that conductive element.

There are other grounding requirements in Article 770. Part II of Article 770, titled Protection, contains only one section, Section 770.33, which is titled Grounding of Entrance Cables. This section requires optical fiber entrance cables that are "exposed to contact with electric light or power conductors" and have "non-current-carrying metallic members" to be grounded at the point of entrance or have the metallic members interrupted and insulated at the point of entrance. The *NEC* actually requires the grounding or interrupting and insulating to be done "as close to the point of entrance as practicable." *Practicable* means "capable of being done" or "can be done in practice" and is similar to but not precisely the same as either "possible" or "practical." Obviously this rule does not apply to nonconductive cables, but does require conductive cables or composite cables with non-current-carrying metal elements to have those elements grounded or interrupted at the point of entrance. This grounding will help keep transient voltages and similar disturbances from entering a building on the optical fiber cables.

SUMMARY

Optical fiber cables are included in the *NEC* because they are often used in conjunction with electrical conductors or equipment, and to promote an orderly implementation of optical fiber technology. Although fiber optics is not an electrical technology as such, the installation and uses of optical fiber cables are very similar to the installation and uses of many electrical systems and circuits. The *NEC* recognizes three types of optical fiber cables: composite cables that also include electrical circuit conductors, conductive cables that include conductive elements that are not intended to carry current, and nonconductive cables that have no metallic members. Cables are listed and marked in accordance with their intended uses, and similar special nonmetallic raceways are listed in similar categories. Generally, optical fiber installations may be done using only the listed cables or using cables in raceways, but some limited uses of unlisted cables are also permitted by the *Code*. Optical fiber installations are similar to installations of limited energy circuits in many ways. In fact, composite cables are actually listed based on the type of electrical circuit conductors included in their construction. Conductive and nonconductive optical fiber cables are subject to specific requirements for grounding and separations from electrical circuits that generally permit the cables to share enclosures with limited energy circuits and prohibit the optical fiber cables from sharing enclosures with higher powered circuits.

Optical Fiber Cables and Raceways

I. General

770.1 Scope. The provisions of this article apply to the installation of optical fiber cables and raceways. This article does not cover the construction of optical fiber cables and raceways.

770.2 Definitions.

Abandoned Optical Fiber Cable. Installed optical fiber cable that is not terminated at equipment other than a connector and not identified for future use with a tag.

Exposed. The circuit is in such a position that, in case of failure of supports and insulation, contact with another circuit may result.

FPN: See Article 100 for two other definitions of *Exposed*.

Optical Fiber Raceway. A raceway designed for enclosing and routing listed optical fiber cables.

Point of Entrance. The point at which the wire or cable emerges from an external wall, from a concrete floor slab, or from a rigid metal conduit or an intermediate metal conduit grounded to an electrode in accordance with 800.40(B).

770.3 Locations and Other Articles. Circuits and equipment shall comply with 770.3(A) and (B). Only those sections of Article 300 referenced in this article shall apply to optical fiber cables and raceways.

(A) Spread of Fire or Products of Combustion. The requirements of 300.21 for electrical installations shall also apply to installations of optical fiber cables and raceways. The accessible portion of abandoned optical fiber cables shall not be permitted to remain.

(B) Ducts, Plenums, and Other Air-Handling Spaces. The requirements of 300.22 for electric wiring shall also apply to installations of optical fiber cables and raceways where they are installed in ducts or plenums or other space used for environmental air.

Source: NFPA 70, *National Electrical Code®*, NFPA, Quincy, MA, 2002 edition.

Exception: As permitted in 770.53(A).

770.4 Optical Fiber Cables. Optical fiber cables transmit light for control, signaling, and communications through an optical fiber.

770.5 Types. Optical fiber cables can be grouped into three types.

(A) Nonconductive. These cables contain no metallic members and no other electrically conductive materials.

(B) Conductive. These cables contain non–current-carrying conductive members such as metallic strength members, metallic vapor barriers, and metallic armor or sheath.

(C) Composite. These cables contain optical fibers and current-carrying electrical conductors, and shall be permitted to contain non–current-carrying conductive members such as metallic strength members and metallic vapor barriers. Composite optical fiber cables shall be classified as electrical cables in accordance with the type of electrical conductors.

770.6 Raceways for Optical Fiber Cables. The raceway shall be of a type permitted in Chapter 3 and installed in accordance with Chapter 3.

Exception: Listed nonmetallic optical fiber raceway identified as general-purpose, riser, or plenum optical fiber raceway in accordance with 770.51 and installed in accordance with 362.24 through 362.56, where the requirements applicable to electrical nonmetallic tubing shall apply. Unlisted underground or outside plant construction plastic innerduct shall be terminated at the point of entrance.

> FPN: For information on listing requirements for optical fiber raceways, see UL 2024, *Standard for Optical Fiber Raceways*.

Where optical fiber cables are installed within the raceway without current-carrying conductors, the raceway fill tables of Chapter 3 and Chapter 9 shall not apply.

Where nonconductive optical fiber cables are installed with electric conductors in a raceway, the raceway fill tables of Chapter 3 and Chapter 9 shall apply.

770.7 Access to Electrical Equipment Behind Panels Designed to Allow Access. Access to electrical equipment shall not be denied by an accumulation of cables that prevents removal of panels, including suspended ceiling panels.

770.8 Mechanical Execution of Work. Optical fiber cables shall be installed in a neat and workmanlike manner. Cables installed exposed on the surface of ceiling and sidewalls shall be supported by the structural components of the building structure in such a manner that the cable will not be damaged by normal building use. Such cables shall be attached to structural components by straps, staples, hangers, or similar fittings designed and installed so as not to damage the cable. The installation shall also conform with 300.4(D).

II. Protection

770.33 Grounding of Entrance Cables. Where exposed to contact with electric light or power conductors, the non–current-carrying metallic members of optical fiber cables entering buildings shall be grounded as close to the point of entrance as practicable or shall be interrupted as close to the point of entrance as practicable by an insulating joint or equivalent device.

III. Cables Within Buildings

770.49 Fire Resistance of Optical Fiber Cables. Optical fiber cables installed as wiring within buildings shall be listed as being resistant to the spread of fire in accordance with 770.50 and 770.51.

770.50 Listing, Marking, and Installation of Optical Fiber Cables. Optical fiber cables in a building shall be listed as being suitable for the purpose, and cables shall be marked in accordance with Table 770.50.

Exception No. 1: Optical fiber cables shall not be required to be listed and marked where the length of the cable within the building, measured from its point of entrance, does not exceed 15 m (50 ft) and the cable enters the building from the outside and is terminated in an enclosure.

FPN: Splice cases or terminal boxes, both metallic and plastic types, are typically used as enclosures for splicing or terminating optical fiber cables.

Exception No. 2: Conductive optical fiber cable shall not be required to be listed and marked where the cable enters the building from the outside and is run in rigid metal conduit or intermediate metal conduit and such conduits are grounded to an electrode in accordance with 800.40(B).

Exception No. 3: Nonconductive optical fiber cables shall not be required to be listed and marked where the cable enters the building from the outside and is run in raceway installed in compliance with Chapter 3.

FPN No. 1: Cable types are listed in descending order of fire resistance rating. Within each fire resistance rating, nonconductive cable is listed first, since it may substitute for the conductive cable.

FPN No. 2: See the referenced sections for requirements and permitted uses.

Table 770.50   Cable Markings

Cable Marking	Type	Reference
OFNP	Nonconductive optical fiber plenum cable	770.51(A) and 770.53(A)
OFCP	Conductive optical fiber plenum cable	770.51(A) and 770.53(A)
OFNR	Nonconductive optical fiber riser cable	770.51(B) and 770.53(B)
OFCR	Conductive optical fiber riser cable	770.51(B) and 770.53(B)
OFNG	Nonconductive optical fiber general-purpose cable	770.51(C) and 770.53(C)
OFCG	Conductive optical fiber general-purpose cable	770.51(C) and 770.53(C)
OFN	Nonconductive optical fiber general-purpose cable	770.51(D) and 770.53(C)
OFC	Conductive optical fiber general-purpose cable	770.51(D) and 770.53(C)

770.51 Listing Requirements for Optical Fiber Cables and Raceways. Optical fiber cables shall be listed in accordance with 770.51(A) through (D), and optical fiber raceways shall be listed in accordance with 770.51(E) through (G).

(A) Types OFNP and OFCP. Types OFNP and OFCP nonconductive and conductive optical fiber plenum cables shall be listed as being suitable for use in ducts, plenums, and other space used for environmental air and shall also be listed as having adequate fire-resistant and low smoke-producing characteristics.

> FPN: One method of defining low smoke-producing cables is by establishing an acceptable value of the smoke produced when tested in accordance with NFPA 262-1999, *Standard Method of Test for Flame Travel and Smoke of Wires and Cables for Use in Air-Handling Spaces*, to a maximum peak optical density of 0.5 and a maximum average optical density of 0.15. Similarly, one method of defining fire-resistant cables is by defining maximum allowable flame travel distance of 1.52 m (5 ft) when tested in accordance with the same test.

(B) Types OFNR and OFCR. Types OFNR and OFCR nonconductive and conductive optical fiber riser cables shall be listed as being suitable for use in a vertical run in a shaft or from floor to floor and shall also be listed as having fire-resistant characteristics capable of preventing the carrying of fire from floor to floor.

> FPN: One method of defining fire-resistant characteristics capable of preventing the carrying of fire from floor to floor is that the cables pass the requirements of ANSI/UL 1666-1997, *Standard Test for Flame Propagation Height of Electrical and Optical-Fiber Cable Installed Vertically in Shafts.*

(C) Types OFNG and OFCG. Types OFNG and OFCG nonconductive and conductive general-purpose optical fiber cables shall be listed as being suitable for general-purpose use, with the exception of risers and plenums, and shall also be listed as being resistant to the spread of fire.

> FPN: One method of defining *resistance to the spread of fire* is for the damage (char length) not to exceed 1.5 m (4 ft 11 in.) when performing the vertical flame test for cables in cable trays, as described in CSA C22.2 No. 0.3-M-1985, *Test Methods for Electrical Wires and Cables.*

(D) Types OFN and OFC. Types OFN and OFC nonconductive and conductive optical fiber cables shall be listed as being suitable for general-purpose use, with the exception of risers, plenums, and other spaces used for environmental air, and shall also be listed as being resistant to the spread of fire.

> FPN: One method of defining *resistant to the spread of fire* is that the cables do not spread fire to the top of the tray in the vertical-tray flame test in ANSI/UL 1581-1991, *Reference Standard for Electrical Wires, Cables, and Flexible Cords.*
>
> Another method of defining *resistant to the spread of fire* is for the damage (char length) not to exceed 1.5 m (4 ft 11 in.) when performing the vertical flame test for cables in cable trays, as described in CSA C22.2 No. 0.3-M-1985, *Test Methods for Electrical Wires and Cables.*

(E) Plenum Optical Fiber Raceway. Plenum optical fiber raceways shall be listed as having adequate fire-resistant and low smoke-producing characteristics.

(F) Riser Optical Fiber Raceway. Riser optical fiber raceways shall be listed as having fire-resistant characteristics capable of preventing the carrying of fire from floor to floor.

(G) General-Purpose Optical Fiber Cable Raceway. General-purpose optical fiber cable raceway shall be listed as being resistant to the spread of fire.

770.52 Installation of Optical Fibers and Electrical Conductors.

(A) With Conductors for Electric Light, Power, Class 1, Non–Power-Limited Fire Alarm, or Medium Power Network-Powered Broadband Communications Circuits. Optical fibers shall be permitted within the same composite cable for electric light, power, Class 1, non–power-limited fire alarm, or medium power network-powered broadband communications circuits operating at 600 volts or less only where the functions of the optical fibers and the electrical conductors are associated.

Nonconductive optical fiber cables shall be permitted to occupy the same cable tray or raceway with conductors for electric light, power, Class 1, non–power-limited fire alarm, or medium power network-powered broadband communications circuits operating at 600 volts or less. Conductive optical fiber cables shall not be permitted to occupy the same cable tray or raceway with conductors for electric light, power, Class 1, non–power-limited fire alarm, or medium power network-powered broadband communications circuits.

Composite optical fiber cables containing only current-carrying conductors for electric light, power, Class 1 circuits rated 600 volts or less shall be permitted to occupy the same cabinet, cable tray, outlet box, panel, raceway, or other termination enclosure with conductors for electric light, power, or Class 1 circuits operating at 600 volts or less.

Nonconductive optical fiber cables shall not be permitted to occupy the same cabinet, outlet box, panel, or similar enclosure housing the electrical terminations of an electric light, power, Class 1, non–power-limited fire alarm, or medium power network-powered broadband communications circuit.

Exception No. 1: Occupancy of the same cabinet, outlet box, panel, or similar enclosure shall be permitted where nonconductive optical fiber cable is functionally associated with the electric light, power, Class 1, non–power-limited fire alarm, or medium power network-powered broadband communications circuit.

Exception No. 2: Occupancy of the same cabinet, outlet box, panel, or similar enclosure shall be permitted where nonconductive optical fiber cables are installed in factory- or field-assembled control centers.

Exception No. 3: In industrial establishments only, where conditions of maintenance and supervision ensure that only qualified persons service the installation, nonconductive optical fiber cables shall be permitted with circuits exceeding 600 volts.

Exception No. 4: In industrial establishments only, where conditions of maintenance and supervision ensure that only qualified persons service the installation, composite optical fiber cables shall be permitted to contain current-carrying conductors operating over 600 volts.

Installations in raceway shall comply with 300.17.

(B) With Other Conductors. Optical fibers shall be permitted in the same cable, and conductive and nonconductive optical fiber cables shall be permitted in the same cable tray, enclosure, or raceway with conductors of any of the following:

(1) Class 2 and Class 3 remote-control, signaling, and power-limited circuits in compliance with Article 725
(2) Power-limited fire alarm systems in compliance with Article 760
(3) Communications circuits in compliance with Article 800
(4) Community antenna television and radio distribution systems in compliance with Article 820
(5) Low-power network-powered broadband communications circuits in compliance with Article 830

(C) Grounding. Non–current-carrying conductive members of optical fiber cables shall be grounded in accordance with Article 250.

770.53 Applications of Listed Optical Fiber Cables and Raceways. Nonconductive and conductive optical fiber cables shall comply with any of the requirements given in 770.53(A) through (E) or where cable substitutions are made as shown in 770.53(F).

(A) Plenum. Cables installed in ducts, plenums, and other spaces used for environmental air shall be Type OFNP or OFCP. Abandoned cables shall not be permitted to remain. Types OFNR, OFCR, OFNG, OFN, OFCG, and OFC cables installed in compliance with 300.22 shall be permitted. Listed plenum optical fiber raceways shall be permitted to be installed in ducts and plenums as described in 300.22(B) and in other spaces used for environmental air as described in 300.22(C). Only types OFNP and OFCP cables shall be permitted to be installed in these raceways.

(B) Riser. Cables installed in risers shall be as described in any of the following:

(1) Cables installed in vertical runs and penetrating more than one floor, or cables installed in vertical runs in a shaft, shall be Type OFNR or OFCR. Floor penetrations requiring Type OFNR or OFCR shall contain only cables suitable for riser or plenum use. Abandoned cables shall not be permitted to remain. Listed riser optical fiber raceways shall be permitted to be installed in vertical riser runs in a shaft from floor to floor. Only Types OFNP, OFCP, OFNR and OFCR cables shall be permitted to be installed in these raceways.
(2) Types OFNG, OFN, OFCG, and OFC cables shall be permitted to be encased in a metal raceway or located in a fireproof shaft having firestops at each floor.

(3) Types OFNG, OFN, OFCG, and OFC cables shall be permitted in one- and two-family dwellings.

FPN: See 300.21 for firestop requirements for floor penetrations.

(C) Other Wiring Within Buildings. Cables installed in building locations other than the locations covered in 770.53(A) and (B) shall be Type OFNG, OFN, OFCG, or OFC. Such cables shall be permitted to be installed in listed general-purpose optical fiber raceways.

(D) Hazardous (Classified) Locations. Cables installed in hazardous (classified) locations shall be any type indicated in Table 770.53.

(E) Cable Trays. Optical fiber cables of the types listed in Table 770.50 shall be permitted to be installed in cable trays.

FPN: It is not the intent to require that these optical fiber cables be listed specifically for use in cable trays.

(F) Cable Substitutions. The substitutions for optical fiber cables listed in Table 770.53 shall be permitted.

Table 770.53   Cable Substitutions

Cable Type	Permitted Substitutions
OFNP	None
OFCP	OFNP
OFNR	OFNP
OFCR	OFNP, OFCP, OFNR
OFNG, OFN	OFNP, OFNR
OFCG, OFC	OFNP, OFCP, OFNR, OFCR, OFNG, OFN

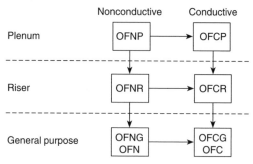

Figure 770.53 Cable substitution hierarchy.

Communications Systems and Circuits

Chapters 3, 4, and 5 of this book each covered primarily one article of the *National Electrical Code®* (*NEC®*) (Articles 725, 760, and 770, respectively). In each case, the use of the types of circuits covered was significantly different in the three articles corresponding to those chapters, although the requirements were similar in many ways. In spite of the similarities in the rules, the three subjects of those articles—remote control and signaling, fire alarm systems and circuits, and fiber optics—are three distinctly separate subjects and three types of systems.

All of the systems in Chapter 8 of the *NEC* are collectively called "communications systems" which is also the title of the chapter. Within Chapter 8, "communications circuits" are the subject of Article 800. When we are talking about CATV (community antenna TV) systems, for example, we are referring to a type of communications system, but its circuits are CATV circuits even though cable types listed for use in communications circuits may be used.

This present chapter covers communications systems generally, or Chapter 8 of the *NEC*. Chapter 8 contains four articles. Article 800 covers communications circuits; Article 810 covers radio and television equipment, primarily antennas; Article 820 covers community antenna television and radio distribution (CATV); and Article 830 covers network-powered broadband communications (NPBC) systems. Although different in some ways, these subjects are closely related to one another.

Communications circuits are traditionally voice circuits, but they are also used for other functions. While cable TV systems fit neatly under Article 820, satellite dish systems tend to fall under Article 810 for the dish antenna and under Article 820 for the interior wiring and cable grounding. Article 830 covers a technology that combines voice, audio, data, and video on one line. It provides requirements for the network wiring up to the network interface unit, where the broadband signal is separated into its various components. According to Section 830.3(D), the output circuits from the network interface unit are covered by Articles 800, 820, 770, 760, and 725, as appropriate for the types of output circuits used. Cable TV systems may be used for video, audio, Internet access, and even voice communications. Satellite dish systems may be used in much the same way as CATV, except that satellite systems typically provide only one-way communications to users. Finally, communications, especially newer versions, such as DSL (digital subscriber line), are now offering broadband applications with multiple uses.

Because of the overlap between some articles, the similarity of the rules in the articles, and the fact that all of these systems may be used in a similar manner, the four articles of Chapter 8 are treated together herein. Actually, little emphasis is placed on Article 810, but selected requirements are discussed to keep all the articles in an appropriate context.

All the systems in Chapter 8 of the *NEC* usually involve a wiring system that enters a building from outside, and those entrance cables are all treated in a similar manner. Sometimes the exterior wiring remains entirely on one site, as in satellite dish systems. In other systems, such as conventional telephone or NPBC, the wiring to a building is an extension of a vast wiring system that is mostly outside. In cases in which control or fire alarm systems are extended outside between buildings, they may be reclassified and treated as communications. At least two characteristics set most of the Chapter 8 systems apart from other wiring systems in buildings. One is that the systems represent an exterior power source (albeit a power-limited source) that has no disconnect as such. Another is that such exterior systems are susceptible to contact with other higher powered systems and therefore require protectors where they enter buildings. Protectors, in turn, are required to be grounded.

The differences between Chapter 8 systems and the other systems covered by the *NEC* are reflected in the organization of the *NEC*. According to Section 90.3, Chapter 8 is independent of the other chapters of the *Code*, and the applicability of those other chapters to Chapter 8 systems is strictly up to Chapter 8. Unlike Articles 725, 760, and 770, Articles 800, 810, 820, and 830 do not exclude themselves from the requirements of Article 300 because Article 300 and all the other articles of Chapters 1 through 7 apply only when the articles in Chapter 8 say so. However, restrictions that are placed on certain wiring methods in Chapter 5 of the *Code* may require that systems be installed using a specific wiring method without regard to the type of system involved. This is especially true in hazardous locations, but the requirements for hazardous location wiring are recognized in the articles of Chapter 8.

The *NEC* does apply to communications circuits unless they are entirely outside of buildings or in spaces under the exclusive control of communications utilities or are otherwise excluded from the rules of the *NEC* by Section 90.2(B).

COMMON REQUIREMENTS

Article 810 primarily covers antennas and lead-in conductors from antennas, so the organization of Article 810 is unique. However, Articles 800, 820, and 830 are organized in a similar manner. There are some differences in the location of some rules, but generally these three articles parallel one another. Each article has the same number of parts with almost identical titles, and within the parts, the section numbers and their subject matter are similar. In this chapter, the fundamental content of each part is covered first, followed by a discussion of some of the specific requirements for each type of communications system.

Part I in each article (including Article 810) contains general rules such as a scope statement, definitions, references to other articles, and general requirements for installations. As previously noted, the locations of rules are not precisely the same in each article. For example, Sections 800.4, 820.3(C), and 830.3(C) each refer to Section 110.3(B) that requires compliance with listing and labeling instructions, but each of these sections is in Part I of the respective article.

Articles 800, 820, and 830 all have a Part II that covers conductors or cables that are outside of or entering buildings. This part in each article covers such things as clearances above roofs, clearances from other conductors (including lightning conductors), and requirements where conductors enter buildings. Article 830 includes clearances above ground and burial depths. Article 810 covers similar issues for lead-in conductors from outside antennas.

Each of the Articles 800, 820, and 830 also has a Part III, titled Protection. Protection in this sense is protection from overcurrents or overvoltages that may be imposed on an outside line due to lightning, contact with or induction from power conductors, or rises in ground potential. Actual requirements for protectors vary between articles. Protectors may be allowed to be fuseless, providing only overvoltage protection, or may be required to be fused to also provide overcurrent protection. In Articles 820 and 830, grounding of conductors may be the only protection required. Article 810 requires a similar type of protection, called "antenna discharge units" on the lead-in conductors in many installations.

Each article in Chapter 8 covers grounding methods. The grounding methods are primarily for grounding entrance cables and raceways, and protectors. Part IV in Articles 800, 820, and 830 cover these requirements. In Article 810, grounding methods are provided in Section 810.21. The actual methods in the four articles are virtually identical in effect, although the wording varies slightly.

Wires and cables or wiring methods within buildings are covered by Part V for communications, CATV, and NPBC circuits in Articles 800, 820, and 830. In these articles, Part V covers cable and conductor types and markings, where each type may be used, and requirements for separations from other circuits. Other than providing clearance and enclosure requirements for antenna conductors for interior transmitting stations, Article 810 does not cover inside wiring requirements. Instead, Section 810.3 refers to Chapters 1 through 4 and Article 640 for audio circuits, and to Article 820 for coaxial cables that are run inside buildings.

As mentioned earlier, Article 810 has its own organization. In addition to Part I, General, Article 810 also has the following parts:

- Part II: Receiving Equipment—Antenna Systems. This part covers the construction, clearance, and grounding requirements for antennas and lead-in conductors, including clearances, and requirements for antenna discharge units and grounding.

- Part III: Amateur Transmitting and Receiving Stations. This part covers similar requirements to those in Part II, but the requirements are modified somewhat for amateur systems. However, Sections 810.11 through 810.15 from Part II also apply to Part III. Section 810.15 covers the grounding requirements for amateur systems.

- Part IV: Interior Installations—Transmitting Stations. This part primarily covers clearances between antenna conductors and other conductors and the requirements for enclosures for indoor transmitters.

DEFINITIONS

Each of the articles in Chapter 8 refers to Article 100 for definitions of terms. However, except for Article 810, each article also provides new definitions of terms that may or may not have other meanings elsewhere in the *NEC* or other definitions in Article 100. For example, Sections 800.2, 820.2, and 830.2 each contain definitions of *exposed* and *Point of Entrance*. (In Section 830.2, the term is "Exposed to Accidental Contact with Electric Light or Power Conductors," but the definition is identical to the definitions of "exposed" in the other articles.)

In Article 100 and as used generally in the *NEC, exposed* may apply to live parts or to wiring methods. Exposed live parts are energized parts that can be inadvertently touched because they are not "suitably guarded, isolated, or insulated." Exposed wiring methods are "on or attached to the surface or behind panels designed to allow access." The term may also be used in other contexts to mean visible or subject to damage. However, in Chapter 8 "exposed" means "The circuit is in such a position that, in case of failure of supports or insulation, contact with another circuit may result."

Point of entrance is not defined in Article 100, but the term is used in Articles 225, 230, and 240 to describe the point where conductors enter a building or other structure by passing through a wall or other defining boundary. When used in regard to communications systems, the point of entrance is "The point at which the wire or cable emerges from an external wall, from a concrete floor slab, or from a rigid metal conduit or an intermediate metal conduit grounded to an electrode in accordance with 800.40(B)." In Articles 820 and 830, the reference to Section 800.40(B) becomes 820.40(B) and 830.40(B), but the rules in those sections are essentially the same. Elsewhere in the *NEC*, except in Article 770, the point of entrance is at the external wall or floor (or roof) boundary, but in Article 770 and in communications, the point of entry can be extended into a building by placing the entering cable in a properly grounded rigid metal conduit (RMC) or intermediate metal conduit (IMC) raceway, as shown in Figure 6.1. The point of entry then becomes the point where the cable emerges from the raceway.

Abandoned cable is another term that is common to Articles 800, 820, and 830. In each article the term is modified by the type of cable covered by that article. For example, the term is "Abandoned Communications Cable" in Article 800, but the definitions are the same. Abandoned cables of whatever type are cables that are "not terminated at both ends at a connector or other equipment and not identified for future use with a tag." Accessible parts of abandoned cables are not permitted to remain.

Premises as defined in Articles 800 and 820, and *premises wiring* as defined in Article 830 refer to the division of responsibility and ownership for communications systems. Communications and CATV wiring define the premises as beginning at the utility–user network point of demarcation. The idea is the same in Article 830, but the

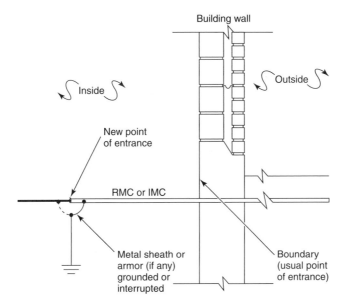

FIGURE 6.1 Point of entrance for communications cable.

dividing point is called the *network interface unit*. The *NEC* is primarily concerned with the wiring that takes place on the user side of the network interface unit or network point of demarcation. Figure 6.2 shows a network point of demarcation on a dwelling unit. This enclosure has two nested doors. The outside door provides access to the user side of the wiring, and an inside door, with tamper-resistant closures, provides restricted access to the protector and the utility side of the wiring.

A *block* is a geographical area bounded by streets, including alleys, but not including the streets. Blocks are important in Articles 800 and 830 because they are used to determine when conductors are required to have protectors. Other factors are also used to determine the need for a primary protector where cables enter buildings, but, generally, conductors contained within a block do not require protectors according to the *NEC*.

Other terms with definitions specific to one article are discussed as they arise in the context of the rules that are unique or specific to that article.

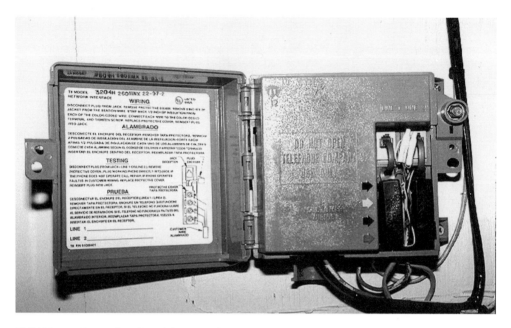

FIGURE 6.2 Network point of demarcation.

POWER LIMITATIONS

The communications systems of Chapter 8 are generally considered to be power-limited circuits. The extent to which a given type of circuit is power-limited depends on the type of circuit.

Article 800 does not specify a voltage or current limit for communications circuits, but the power limits are well established by practice and product standards in the communications industry. Common telephone circuits are powered by the communications utility, usually at an open-circuit voltage of about 48 volts DC. The voltage includes an AC component and fluctuates during use, going higher when a phone is ringing and lower when the phone is off the hook and in use. Current limitations are established by the communications circuitry or, where necessary, by a fused primary protector.

The antenna systems covered by Article 810 do not have voltage or current limits specified in the *NEC*. However, these circuits are inherently energy-limited by the fact that the currents are induced in the antennas by broadcast radio frequencies. These relatively low-energy signals must be amplified to produce enough power to drive audio equipment. Even after amplification, the most common types of audio circuits are still considered to be energy-limited to the point that they are usually permitted under Article 640 to use Class 2 and Class 3 wiring methods.

As covered by Article 820, CATV circuits and similar circuits do have specific voltage limitations. The coaxial cables used under Article 820 are permitted to be used only on systems of up to 60 volts. Section 820.4, in which the 60 volt limit is found, also requires the current to be supplied from an energy-limiting source.

Two types of NPBC systems are recognized by the *Code*. The types are based on power levels. Low-power NPBC systems are limited to 100 volts and 250 volt-amperes. Medium-power systems are limited to 150 volts and 250 volt-amperes. Nameplate ratings for power supplies for both power levels are limited to 100 volt-amperes. Like Class 2 and Class 3 circuits, the power sources of NPBC systems are required to be inherently power-limited, either by a current-limiting impedance or a combination of a current-limiting impedance and an overcurrent device. These power limitations are found in Section 830.4 and Table 830.4, which is reproduced as Figure 6.3. The power limitations for NPBC systems are comparable to the limitations for Class 3 and PLFA (power-limited fire alarm) circuits.

As suggested by the use of low- and medium-powered systems, the original proposal to add Article 830 to the *NEC* also included a high-power classification in which nameplate ratings for power supplies could have been as high as 2,250 volt-amperes or about equivalent to a 20 ampere, 120 volt branch circuit. However, the high-power classification was not accepted into the *NEC*.

ABANDONED CABLES

The concern with abandoned cables is that they represent a quantity of combustible material with no purpose. Such cables add no functionality but do add to the possible fire load. Therefore, Articles 800, 820, and 830 require abandoned cables without raceways to be

Network Power Source	Low	Medium
Circuit voltage, V_{max} (volts)[1]	0–100	0–150
Power limitation, VA_{max} (volt-amperes)[1]	250	250
Current limitation, I_{max} (amperes)[1]	$1000/V_{max}$	$1000/V_{max}$
Maximum power rating (volt-amperes)	100	100
Maximum voltage rating (volts)	100	150
Maximum overcurrent protection (amperes)[2]	$100/V_{max}$	NA

[1]V_{max}, I_{max}, and VA_{max} are determined with the current-limiting impedance in the circuit (not bypassed) as follows:

V_{max}—Maximum system voltage regardless of load with rated input applied.

I_{max}—Maximum system current under any noncapacitive load, including short circuit, and with overcurrent protection bypassed if used.

I_{max} limits apply after 1 minute of operation.

VA_{max}—Maximum volt-ampere output after 1 minute of operation regardless of load and overcurrent protection bypassed if used.

[2]Overcurrent protection is not required where the current-limiting device provides equivalent current limitation and the current-limiting device does not reset until power or the load is removed.

FIGURE 6.3   Limitations for network-powered broadband communications systems. (Source: *National Electrical Code®*, NFPA, 2002, Table 830.4)

removed from plenums, risers, and hollow spaces. These locations are the areas in which unnecessary additions to the fire load are most undesirable. Cables that are stapled in place in concealed locations or installed in *NEC* Chapter 3 raceways are not required to be removed. For example, in Article 800, a requirement to remove abandoned communications cables is found in Section 800.53(A) regarding cables in plenums and Section 800.53(B)(1) covering cables in risers. However, the requirement to remove the abandoned cables is not found in 800.53(B)(2) or 800.53(B)(3) where the cables are installed in metal raceways or where the cables are installed in one- and two-family dwellings. According to these sections, only abandoned cables in plenums, including other space for environmental air, and in vertical runs without metal raceways must be removed.

Section 820.53 is similar to Section 800.53 because coaxial CATV cables must be removed from plenums according to Section 820.53(A) and from vertical runs without metal raceways according to Sections 820.53(B)(1) and (B)(2). Section 820.53 differs in Section 820.53(D) because there it says "Abandoned cables in hollow spaces shall not be permitted to remain." Section 830.55, which covers low-power NPBC cables, is also essentially the same as 800.53 except for numbering. Section 830.54, which covers medium-power NPBC cables does not mention the removal of abandoned cables. However, in both Sections 820.3(A) and 830.3(A), in which the requirements of Section 300.21 are referenced, "the accessible portion" of abandoned cables are required to be removed. This same language also appears in Section 800.52(B). Therefore, whether the circuits are in communications, CATV, or NPBC cables, it is the accessible portion of abandoned cables that must be removed.

What is meant by the accessible portion of an abandoned cable? Do cables in raceways have to be removed because they are accessible at outlets, pull points, and so on? The definition of *accessible* as applied to wiring methods answers part of these questions. The definitions in Article 100 apply to all four articles because Article 100 is referenced in Section 2 of each article (800.2, 810.2, 820.2, and 830.2). According to the definition in Article 100, "Accessible (as applied to wiring methods). Capable of being removed or exposed without damaging the building structure or finish or not permanently closed in by the structure or finish of the building."

To get the rest of the answer concerning accessibility, we need to consider the opposites of accessible, which would be "inaccessible" or "concealed." Article 100 includes the following definition of *concealed:* "Concealed. Rendered inaccessible by the structure or finish of the building. Wires in concealed raceways are considered concealed, even though they may become accessible by withdrawing them."

These definitions should clarify what cables are intended to be removed, but where there is still some question, the reader should also keep in mind what the intent of the Code Making Panel (CMP) was in accepting these requirements for the 2002 *NEC.* According to CMP 16, ". . . the removal of abandoned cables addresses a significant fire safety issue." It also said that "The intent is not to remove cables where it would be extremely difficult or damaging to the building or the remaining cables." For instance, a raceway might contain numerous cables, and if one cable is abandoned it might be difficult or damaging to the remaining cables to remove the unused cable. As long as the cable is within the raceway, the presence of the abandoned cable itself is not a fire hazard because it is isolated from other combustibles by the raceway. In this case, according to the definitions, conductors in a raceway are not considered to be accessible anyway.

In the original proposal to require removal of abandoned cables, the substantiation for the proposal said that if the *NEC* required removal of abandoned cables, the removal would become a part of the specification and contract for a new communications installations. Certainly installers should now remove any cables that are abandoned as a result of new installations, and building, fire, or insurance inspectors have a basis in the *NEC* to demand that the cables are removed.

SPREAD OF FIRE AND PRODUCTS OF COMBUSTION

Abandoned cables are part of the larger issue that is concerned with an electrical installation contributing to the spread of fire or smoke. Electrical installations may be a concern due to either the electrical materials being combustible or to the electrical installation producing penetrations and openings in fire-rated construction assemblies. Any penetrations in fire-rated assemblies have to be restored to a fire-rated condition, and combustible materials may be limited or prohibited in certain spaces. Limited energy circuits are not likely to start fires.

In Article 820, Section 820.3(A), the requirements of Section 300.21 to limit the spread of fire or smoke are incorporated by a direct reference to that section. In Article 800, Section 800.52(B), the language of Section 300.21 is repeated, almost word for word, to provide the same requirements. In Article 830, a direct reference is made in Section 830.3(A), *and* the language is reiterated in Section 830.58(B) for low-power NPBC circuits.

The specific requirements for limiting wiring in air-handling spaces are closely related to the issues of fire spread and smoke production. These requirements are incorporated by a reference to Section 300.22 in Sections 820.3(B) and 830.3(B), and again in Sections 820.53(A) and 830.55(B) for CATV and NPBC conductors. In Article 800, Section 800.53(A) refers to Section 300.22. Note that in each case, the references to Section 300.22 effectively prohibit the use of any wiring method in ducts that handle loose stock or vapor. Also, plenum-rated cables, plenum-rated communications raceways (for communications circuits), metallic raceways, or metallic-sheathed cables are required in other ducts, plenums, and air-handling spaces.

OUTSIDE CABLES

The basis for a concern with conductors that are outside buildings is simply that they will eventually enter buildings. The conductors of the limited energy systems covered by Chapter 8 in the *NEC* do not normally present significant fire or shock hazards in and of themselves. However, cables that are installed outside of buildings may be run on utility poles, in manholes, or underground in trenches that are shared by other circuits. These conditions make the circuits covered by Chapter 8 that are outside buildings more susceptible to contact with higher power circuits and lightning than circuits installed inside buildings. Such unintentional contacts could override the energy-limiting means of the limited energy systems and their safe use would then be seriously compromised. Therefore, Part II of Articles 800, 820, and 830 provides requirements for separation from other circuits and systems to minimize the possibility of contact, as well as requirements intended to protect conductors from damage.

The actual requirements for outside conductors are quite similar in Articles 800, 820, and 830, and the issue is covered in Article 810 as well. For example, Sections 800.10(A), 820.10(A), and 830.11(A) all prohibit the outside conductors covered by the respective articles from being mounted on the same crossarm on a pole with electric lighting or power or similar higher powered conductors. These sections also require their covered conductors to be located below the power and lighting conductors wherever "practicable, " which means wherever it can be done in actual practice, not just where it is "practical" or everywhere it's theoretically "possible." (Although a thesaurus may list "practical" and "possible" as synonyms, the dictionary definitions of these three terms are different.) Figure 6.4 illustrates a common arrangement of different services on a pole.

Article 810 prohibits outside antennas and lead-in conductors from passing above open electric light and power conductors. Also, according to Section 810.13, antennas and lead-in conductors must, where practicable, be kept from passing under open electric light and power conductors.

Part II of Articles 800, 820, and 830 also provides minimum clearances above roofs, and Section 830.11 provides for clearances of overhead NPBC conductors above ground and above pools. Generally, the minimum clearance above a roof is 8 feet, but this distance

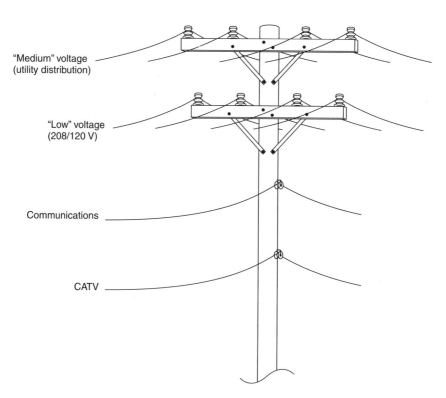

"Medium" voltage
(utility distribution)

"Low" voltage
(208/120 V)

Communications

CATV

FIGURE 6.4 Arrangement of different systems on a pole.

is reduced in specific situations, in much the same way clearances are reduced for service or outside feeder conductors.

CABLES ENTERING BUILDINGS

Concerns with cables entering buildings are related to the possibility of exposure of limited energy circuits to higher energy circuits. These issues are covered by Sections 800.11 through 800.13 for communications, Sections 820.10 and 820.11 for CATV, and Sections 830.11 and 830.12 for NPBC circuits. In the case of communications circuits, the location of the protector is also a consideration.

Communications, CATV, and NPBC cables on buildings must be securely attached to maintain at least 4 inches physical separation from non-power-limited circuits, or must be separated by continuous fixed nonconductors. They should also be separated at least 6 feet from lightning conductors. Where CATV or NPBC cables enter a building from underground, they must be separated from non-power-limited circuits by separate raceways, by permanent barriers, or, if directly buried, by a 12 inch separation. Conductors of separate systems are considered to be separated if either the power-limited conductors or the non-power-limited conductors are installed in metal-sheathed or metal-clad cables or armor, or if the non-power-limited conductors are installed in Type UF or USE cables.

Installations of communications cables on buildings must maintain separations from higher power conductors. In the case of communications circuits, the location of the protector determines what entry methods may be used. If the protector is outside so that the entering cables are protected, only the separations discussed later in this chapter under "Circuit Separations" are required. If the protector is inside the building, then the entering conductors are not protected, and bushings or raceways or metal-clad cables are required where the conductors pass through the wall unless the wall is masonry. Metal raceways used to contain unprotected communications cables must be grounded.

PROTECTION AND PROTECTORS

The systems covered by Chapter 8 usually involve a system that comes from outside a building. The conductors of such systems are often *exposed* as defined in Chapter 8, that is, subject to contact with higher power systems or lightning. Such contact or induced currents may result in overvoltages or overcurrents on the circuit conductors. Protection is required to reduce the likelihood that these possible overvoltages or overcurrents will be conducted into buildings. For antenna systems, Section 810.20 requires an antenna discharge unit unless the lead-in conductors have a continuous grounded metallic shield. Similarly, Article 820 requires grounding of a coaxial cable shield as near as practicable to the point where the CATV conductors enter the building. For antennas, CATV, and NPBC circuits, the only protection required may be grounding, but in Article 800, primary protectors are required for exposed circuits. Article 830 requires primary protectors on NPBC conductors that are exposed and not grounded or interrupted.

Underwriters Laboratory Standard UL 497, *Standard for Safety for Protectors for Paired-Conductor Communications Circuits,* covers the primary protectors required by Article 800. According to this standard, a protector is "intended to protect equipment, wiring, and personnel against the effects of excessive potentials and currents in telephone lines caused by lightning, contacts with power conductors, power induction, and rises in ground potential." Fuseless protectors consist of an arrestor, which is a device that blocks current normally, but conducts current, usually to ground, when a predetermined voltage level is reached. In addition to the arrestor, protectors include an insulating mounting base and provisions for the connection of telephone and grounding conductors. Fuses are included along with the arrestors to form fused primary protectors. Figure 6.5 is a photo of a primary protector of the type commonly used in commercial telecommunications installations. This protector actually has provisions for multiple pairs and multiple protectors as well as terminals for grounding conductors.

Fuseless protectors may be used when fault currents from power systems are limited to less than the current rating of the protector

FIGURE 6.5 Primary protector. (Source: *National Electrical Code® Handbook,* NFPA, 1999, Exhibit 800.3; courtesy of AT&T)

and the current capacity of the grounding conductors. Where available fault currents at a protector are greater than the current-carrying capacity of the protector or grounding conductors, a fused protector must be used to limit currents to levels that are safe for interior wiring.

Standard UL 497 details many construction requirements and test procedures for primary protectors. Protectors are frequently connected to circuits that operate at about 48 volts. Such protectors are required to withstand 1,000 volts AC rms for 1 minute without breakdown under a specified procedure in which the voltage is increased slowly. However, when the voltage is increased rapidly, the arrestor must break down at a voltage within 25 percent of its rated value, and at no more than 750 volts. The communications circuits that are protected are typically in communications cables. Communications conductors in cables are rated at 300 volts, so protectors are usually selected to limit transient voltages to about 300 volts. (Listed communications conductors are tested to withstand 1,500 volts AC or 2,500 volts DC.)

Secondary protectors are also available and are permitted but not required on communications circuits. However, secondary protectors may not be used in lieu of primary protectors, so they may be used only in conjunction with primary protectors or where primary protectors are not required. Secondary protectors for communications circuits are sometimes provided with transient voltage surge suppressors (TVSS) used for power line protection, and they are listed under UL 497A, *Standard for Safety for Secondary Protectors for Communications Circuits*. The UL standard for TVSS devices is UL 1449, *Standard for Transient Voltage Surge Suppressors*. Some equipment may be listed under both of these standards and possibly others. (TVSS devices for power circuits are covered by Article 285 of the *Code*.)

Telephone utilities commonly furnish a protector for each circuit pair at the point of entry to a building. Some of these protectors may not be required by the *NEC*, if, for example, the circuit is not exposed and remains within a block. However, the *NEC* considers most overhead interbuilding circuits to be exposed to lightning, and additional protectors may be required. In fact, if an interbuilding circuit is exposed to lightning, protectors are required at each end of such circuits. Fine Print Notes in Sections 800.30(A) and 830.30(A) point out that the functions of a primary protector are often beneficial even where the protector is not actually required by the *Code*.

Although the *NEC* does not require primary protectors in all cases, primary protectors and other surge suppressing devices are required for all electrical services when the building is equipped with a lightning protection system. In its Section 3-18, NFPA 780, *Standard for the Installation of Lightning Protection Systems*, reads as follows: "Surge Suppression. Devices suitable for the protection of the structure shall be installed on electric and telephone service entrances and on radio and television antenna lead ins." Certain TVSS devices can provide this protection for power, telephone, and coaxial cable lead-ins in one location, or separate devices may be used. One location may be preferable because it provides a single common grounding reference point for all the protected systems. Figure 6.6 shows an example of a TVSS with secondary protectors.

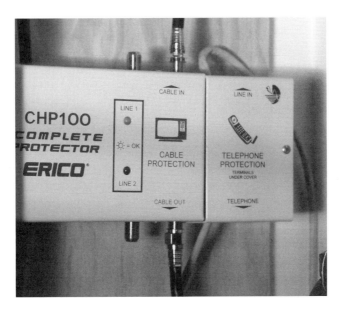

FIGURE 6.6 TVSS with secondary protection for communications and CATV cables.

GROUNDING

Grounding methods are essentially the same in all four articles of Chapter 8 although there are some minor differences. For instance, the minimum size of the grounding conductor is 14 AWG in Articles 800, 820, and 830, but in Article 810, the minimum size is 10 AWG. However, for CATV and NPBC systems, the grounding conductor must also be approximately equal in current-carrying capacity to the coaxial cable shield conductor in the case of CATV or the shield or protected conductors in an NPBC cable. The grounding conductors for connection to an electrode are not required to be larger than 6 AWG.

Grounding conductors should be as short and straight as practicable to limit the difference of potential that might develop across the conductor, and to increase the likelihood that transient currents such as might be caused by lightning will actually follow the conductor to ground. The grounding conductors are required to be connected to a grounding electrode, which generally is either a power system electrode or an electrode that is bonded to the power system electrode system. Once the electrodes are bonded together, they are considered for the purposes of Article 250 to be all the same electrode system. This provision is found in Section 250.58, which requires separate services in a building to use the same electrode.

Regardless of the type of communications system introduced to a building, the choices of an electrode for grounding the system are basically the same. The choices are found in Sections 800.40(B)(1), 810.21(F)(1), 820.40(B)(1), or 830.40(B)(1), depending on the type of system. The choices include the nearest accessible point of any of the following:

- The grounding electrode system
- The grounded interior metal piping
- The outside of power service enclosures
- The metallic power service raceway
- The service equipment enclosure
- The grounding electrode or grounding electrode conductor enclosure
- The grounding conductor or electrode at a separate building disconnecting means

Other options are available when there is no power system, but those situations are highly unusual.

Where a power system exists in a building, the choices of electrodes for communications systems are portions of, or are bonded to, the power grounding electrode system. Most buildings using circuits from Chapter 8 also have power systems. Therefore, in most cases, a connection to one of the choices of electrodes listed results in the bonding of the Chapter 8 system to the power grounding electrode system. Figure 6.7 shows a proper connection to a service raceway. Figure 6.8 shows an improper connection to a branch-circuit raceway. The connections are made with appropriate devices in both cases, but the choice of an electrode in Figure 6.8 is a *Code* violation.

FIGURE 6.7   Correct communications ground connection.

FIGURE 6.8   Incorrect communications ground connection.

Grounding conductors are required to have mechanical protection when subject to physical damage. Where such protection is provided by a metal raceway, the raceway must be bonded to the conductor at each end of the raceway, in the same manner as required for power system grounding electrode conductors. Section 810.21(D) permits oversizing of the grounding conductor in lieu of providing mechanical protection.

Although grounding methods are almost identical from article to article in Chapter 8, the things that have to be grounded are a little more varied. Article 800 requires metallic entrance conduits ahead of protectors and all protectors to be grounded. Metallic cable sheaths may either be interrupted and insulated or grounded at their point of entrance. Masts and other metal structures that support antennas are required to be grounded in Article 810. Article 810 also requires grounding of antenna discharge units where the units are required, and the grounding of a continuous metal shield otherwise. Article 820 requires the grounding of the outer conductive shield of a coaxial cable. Other protective devices are not required in Article 820, but may be used and grounded. In both Articles 820 and 830, RMC or IMC entrance conduits must be grounded as shown in Figure 6.9 if they are used to extend the point of entrance to a point inside the building. In Article 830, metallic cable shields must be grounded at the point of entrance if the shield is used as a circuit

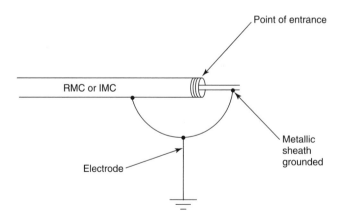

FIGURE 6.9 Grounding of entrance cable and raceway.

conductor, or may be interrupted and insulated if not used as a circuit conductor.

Notice that the requirement for grounding of an entrance conduit in Articles 800, 820, and 830 comes from the definition of "point of entrance," and not from any specific requirement for grounding raceways. Neither RMC nor IMC conduits are required, and other conduits may be used, but only grounded RMC or IMC can extend the point of entrance into a building.

As an example of the interplay between the articles of Chapter 8, consider the installation of a satellite dish antenna for satellite TV reception. The antenna is covered by Article 810 according to Section 810.1. However, the coaxial cable that enters the building is, according to Section 810.3, covered by Article 820. Section 810.20 requires an antenna discharge unit for the lead-in wires unless "the lead-in conductors are enclosed in a continuous metallic shield that is either permanently grounded or protected by an antenna discharge unit." The coaxial cable covered by Article 820 has just such a shield. Section 820.33 requires that shield to be grounded, and where this requirement is met, no other protective device is required. The grounding location in Article 810 is outside or nearest the point of entrance, and the grounding location in Article 820 is nearest the point of entrance. Therefore, a grounding block, located at the point of entrance of the coaxial cable, and properly connected to an electrode as shown in Figure 6.10, meets the requirements of both articles. (Figure 6.10 shows the conductor to the electrode but not the connection to the electrode.) In fact, UL listings require instructions to this effect to be included with the satellite dish and receiver.

Other metal boxes and raceways, such as those used for sleeves between floors or stub-ups from concealed boxes to accessible ceilings as shown in Figure 6.11 are not required to be grounded. Article 250 does not include rules for grounding communications systems since they are covered by Chapter 8, and Chapter 8 does not refer to Article 250 for grounding of raceways other than RMC and IMC entrance conduits. Also, communications circuits usually do not include an equipment grounding conductor in the cable, nor is there an overcurrent device on which to base the sizing of an equipment grounding conductor.

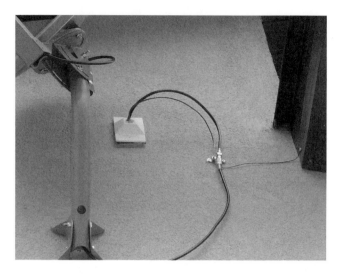

FIGURE 6.10 CATV cable grounding. Grounding block is connected to the electrode and bonded to the metal fascia.

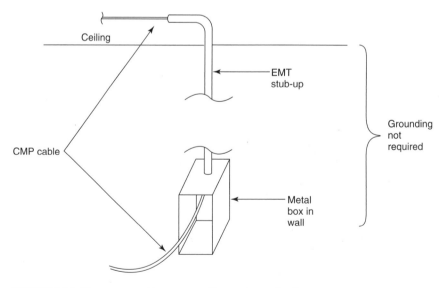

FIGURE 6.11 Raceway and box grounding not required.

WIRING METHODS AND CABLE TYPES

Antennas and lead-in conductors are the focus of Article 810, and other articles are referred to for wiring methods. Special wiring methods are specified for communications, CATV, and NPBC circuits. The special cable types are described in Sections 800.51, 820.51, and 830.5. The nomenclature for the cables differs from one article to another, but the uses and substitution hierarchies are similar.

The basic cable type for communications is Type CM (communications), and the basic type for CATV is Type CATV. There are two types of NPBC cables: Type BM for medium power and Type BL for low power. The basic types are all suitable for general use. General use does not include use in plenums, ducts, risers, or other environmental air-handling spaces unless the cables are installed in metallic raceways as permitted in Section 300.22. Each of the basic or general-use types may have additional letters added to the type to indicate suitability for specific uses. Generally, the suffix P stands for "plenum," the letter R stands for "riser, " and the letter X indicates "limited use." In addition, NPBC cables may also use the letter U to stand for "underground," and communications cables may use the letters UC for "under carpet."

A P suffix indicates that the cable has been tested and passes a plenum flame test that measures flame propagation and smoke density. To be suitable for a plenum application, such cables should not spread a fire rapidly or produce a lot of smoke that would be transported by the air-handling system. This marking makes selection of cables easy for installers and identification of cables to verify compliance easy for inspectors.

An R suffix means that the cable passes a riser flame test. To be installed in a riser, a cable should be slow in propagating flames so that a fire will not spread from floor to floor by following the cable. Riser cables are required where cables are installed in vertical runs in shafts or penetrating more than one floor, unless the cables are installed in metal raceways or in fireproof shafts with firestops at each floor. Riser applications in one- and two-family dwellings do not require riser cables.

Using the marking scheme described, we see that Type CMP, CATVP, and BLP can be used in ducts, plenums, or other space for environmental air without enclosing the cables in a raceway. However, as noted previously, each article also refers to Section 300.22, and Section 300.22(A) says no such wiring is permitted in ducts that handle loose stock or vapor, with or without raceways. Keep in mind that these rules are in Chapter 8, and the rules in Chapter 8 do not modify Section 300.22 or anything in the first four chapters. Chapter 8 stands by itself, and Section 300.22 applies because it is specifically referenced.

Note that there is no Type BMP. Medium-power NPBC circuits installed in plenums or other air-handling spaces must be installed in accordance with Sections 300.22(B) and (C), which require metallic raceways or metallic cable armor.

Vertical runs or riser applications may use Types CMR, CATVR, BMR, or BLR, depending on the circuit type. Alternatively, nonriser types may be used in metal raceways or in fireproof shafts with firestops at each floor. Riser applications in one- and two-family dwellings may utilize either general-purpose or limited-use (suffix X) cables.

Listed nonmetallic communications raceways are permitted to be used in plenum and riser applications as well. The raceways are available in plenum, riser, and general-purpose types. Where a plenum raceway is used in a plenum, however, it may still only contain plenum-rated cables, and when a riser raceway is used in a riser application, it may only contain riser- or plenum-rated cables. These rules are illustrated in Figure 6.12.

Figure 6.13 illustrates and summarizes the permitted applications of communications raceways and communications and multipurpose (Type MP) cables in buildings.

The limited-use Types CMX, CATVX, and BLX (there is no BMX) are intended primarily for use in dwelling units; however, they may also be used in raceways and in limited lengths in nonconcealed spaces.

Other special-purpose types are Type CMUC, which is a communications cable intended for installations under carpets, and Types BMU and BLU, which are intended for outdoor use under-

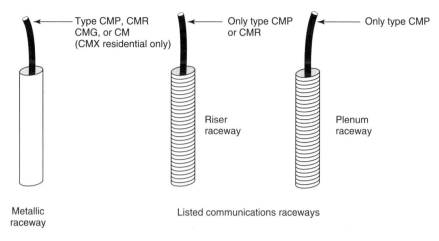

FIGURE 6.12   Use of raceways.

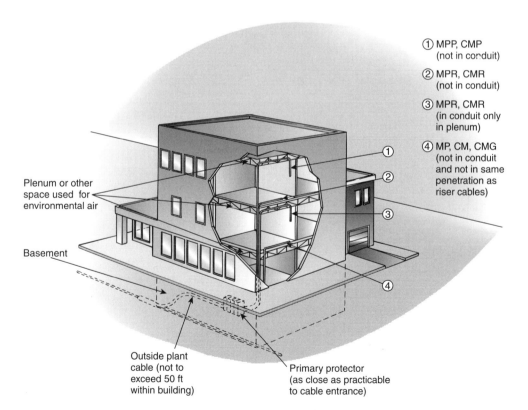

FIGURE 6.13   Applications of communications and multipurpose cables. (Source: *National Electrical Code® Handbook,* **NFPA, 2002, Exhibit 800.3)**

ground only. However, like unlisted communications cables, both Type BMU and Type BLU may be extended into a building if enclosed in a grounded IMC or RMC raceway.

Communications cables (Type CM, CMX, CMR, and CMP) are generally permitted to substitute for the other cable types of Chapter 8, except that the only substitute for Type BM is Type BMR and there is no substitute for type BMR. (Riser applications do not have to use Type BMR; they may use Type BM in metal raceways or fireproof shafts.) Type MP (multipurpose) cables may be substituted for Type CM cables. Substitutions must also take into account the suffix letter if the cables are to be used for a special purpose without raceways. Substitution hierarchies are given in Tables 800.53, 820.53, and 830.58. Articles 800 and 820 also include illustrations of the permitted substitutions, which are reproduced here as Figure 6.14. The hierarchy of substitutions for CATV cables does not include mention of NPBC methods. Nevertheless, Section 820.3(G) permits the use of Article 830 methods in lieu of Article 820 methods, so the BM and BL cable types may also be substituted for the CATV types.

Class 2 cables are at the bottom of the substitution hierarchy, and communications or multipurpose cables are at the top. Class 2 cables are rated for 150 volts, but Class 3 and higher cables are all rated for 300 volts and all undergo similar tests for dielectric strength, insulation crush resistance, jacket peel back, and so on. However, the communications cable testing is a little more demanding, and Type CM is therefore considered to be a better cable. In fact, many cables carry multiple listings and markings so that either through the multiple listings or through the substitution hierarchy they can have multiple applications. For example, Type CL2 cable is sometimes also listed as Type CM, and Type CATV cable may also be listed as Type CL2 or Type CM.

Conduit fill is an issue relating to all types of circuits in Chapter 8. The primary reference that requires compliance with the conduit fill limits of Table 1 in Chapter 9 is in Section 300.17. Article 300 does not apply at all to Chapter 8 unless sections are referenced in Chapter 8. Section 300.17 is not incorporated by any references in

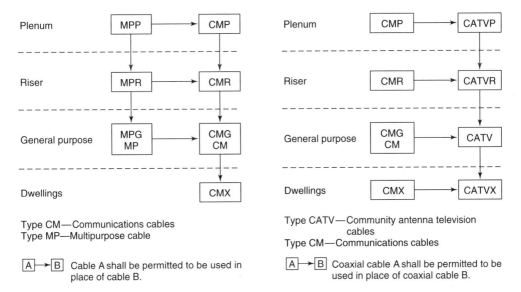

FIGURE 6.14 Substitution hierarchies. (Source: *National Electrical Code®*, NFPA, Figures 800.53 and 820.53)

Chapter 8. In fact, Section 800.48 says that although the general requirements of Chapter 3 apply to raceways installed for communications, conduit fill restrictions do not apply. Articles 820 and 830 simply do not mention the issue. However, the lack of a reference to conduit fill does not necessarily settle the issue.

Here is an example to illustrate the sides to this issue. Suppose we decide to use some RMC for an NPBC cable entry to a building using Type BLU cable. Article 830 says this conduit has to be grounded, but it does not say anything else about the installation of the conduit. To many people, the choice of the conduit brings with it the rules for that wiring method. Since Article 830 does not say anything about it, Article 344 must apply. Section 344.22 says cables are required to meet the requirements of Table 1 in Chapter 9. (A similar rule is found in every other raceway article.) Therefore, it appears that conduit fill limits do apply to NPBC circuit cables.

CIRCUIT SEPARATIONS

Chapter 8 is set aside from the rest of the *NEC* and special rules apply because the circuits it covers have special uses and are considered to be limited energy circuits. The special uses demand special cable types. For example, because of the high frequencies employed in CATV and other broadband technologies, coaxial cables or cables with carefully designed and manufactured twisted pairs are needed. However, as Chapter 8 is set apart from the rest of the *Code*, the wiring methods of Chapter 8 must also be set apart and restrictions are placed on the installation and use of the power-limited cables specified in Chapter 8.

The circuits covered by Chapter 8 get special treatment because they are supplied from power-limited sources. In general, these sources are carefully specified, designed, and tested to be sure that the energy limits are reliable. To be certain that the energy limits will not be compromised, circuits with energy limits must be separated from higher powered circuits. This separation provides reasonable assurance that the power-limited circuits will remain power-limited circuits.

The basic separation requirements for communications, CATV, and NPBC cables are found in Sections 800.52, 820.52, and 830.58. To understand the separation requirements it may be helpful to recognize the things that are restricted, that is, what is required to be separated from what. The various circuit types of Chapter 8 can be divided into three groups: a group that can share the same cable, a group that can share the same raceway or other enclosure, and a group that cannot share the same cable, raceway, or enclosure with the other groups. Also included in these groups are some circuit types from Chapter 7. A fourth group is the group that is comprised of non-power-limited circuits that are not covered by Chapter 8: Class 1 circuits, NPLFA (non-power-limited fire alarm) circuits, and electric lighting and power circuits. Certain restrictions are placed on the intermixing of some of these non-power-limited circuits as well. For example, Class 1 and NPLFA circuits are permitted to share a raceway or enclosure with power circuits only when they are functionally associated or connect to the same equipment. These restrictions are covered in Chapters 3 and 4 of this book.

Group one, the group that can share the same cable, is comprised of communications, Class 2, and Class 3 circuits. Although PLFA circuits can share the same cable with Class 2 or Class 3 circuits, PLFA circuits are not permitted to share a cable with communications circuits. The circuits that can share cables reflect actual likely applications such as data and phone (Class 2 and communications) or controls and fire alarm (Class 2 and PLFA). Sharing a cable in this sense means that the conductors of the various circuits are not individually jacketed within an overall jacket. Figure 6.15 illustrates the circuits that are allowed to share a CM cable.

Group two, the group that can share the same raceway or other enclosure, consists of separate jacketed cables containing Class 2, Class 3, communications, PLFA, CATV, or low-power NPBC circuits, or optical fibers. These cables are permitted to be mixed with other cables in the list, but individual conductors or conductor pairs are not permitted to share the same cable as individual conductors of other circuit types. Listed composite cables may include more than one of these circuit types in a single cable, but such cables are usually constructed like cables within cables. For example, a coaxial member or optical fibers (or both) may be included in a cable that is listed as a communications cable. Section 820.52 does not specifically permit CATV cables to be in the same cable with other circuits, but coaxial members may be included in listed communications cables, and communications cables may be used as substitutes for CATV cables. (Coaxial members may also be included in other listed power-limited cables such as Class 2 cables.)

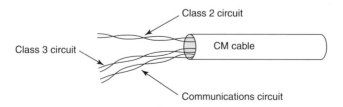

FIGURE 6.15 Circuits occupying the same CM cable.

Group three consists of only one type: medium-power NPBC circuits. Cables containing circuits of this type may not be intermixed in the same raceway or enclosure (or cable) with either the power-limited circuits of group two, or the non-power-limited circuits in the fourth group.

In some cases, limited energy circuits may have to enter or pass through an enclosure that also contains non-power-limited conductors. If the power-limited and non-power-limited conductors must connect to the same equipment, they may share a common enclosure where the common equipment is located and the required separation is permitted to be satisfied by maintaining a ¼ inch physical separation between the power-limited and non-power-limited circuits. The separation requirement may also be met by providing a barrier between the power-limited and non-power-limited conductors in an enclosure.

In addition to the restrictions on mixing of circuits in common enclosures, separations are required to be maintained between power-limited conductors and other conductors outside of enclosures and raceways. Generally, 2 inches of separation must be maintained between the power-limited and non-power-limited circuit conductors. However, the 2 inch spacing may be reduced by enclosing the power-limited conductors in a raceway, or by enclosing the non-power-limited conductors in a raceway or a wiring method with a metallic or nonmetallic sheath, such as Type MC, Type AC, Type NM, or Type UF cable. Alternatively, firmly fixed nonconductors such as flexible tubing may be used to maintain a physical separation and avoid contact between different systems.

SUPPORTS AND CABLE PROTECTION

The support requirements for cables are identical for communications, CATV, and NPBC cables. Sections 800.6, 820.6, and 830.7 all require that installations be done in a "neat and workmanlike manner." This standard is rather subjective. However, the same sections also require that cables exposed on ceiling or wall surfaces be supported by structural components of a building. ("Exposed" in this sense means visible, it does not mean exposed to contact with higher powered systems as defined in Section 800.2, or exposed as applied to wiring methods as defined in Article 100, or exposed as applied to live parts as defined in Article 100.) Fastening means must securely support but not damage the cables. No specific spacing requirements between supports are provided as they are for cables and raceways in Chapter 3 of the *Code*. The installation must be done in such a way that the cable will not be damaged due to normal use of the building. Cables may not be supported by attaching them to raceways, except where the raceway is used as an overhead mast specifically intended for the support and attachment of overhead spans of specific cables. (A mast for a power service may not be used to support other cables.)

Additional installation guidelines and standards are available from other standards organizations, including the Institute of Electrical and Electronics Engineers (IEEE), American National Standards Institute (ANSI), Electronic Industries Alliance (EIA), Telecommunications Industry Association (TIA), and Building Industry Consulting Service International (BICSI). However, these other standards are not requirements of the *NEC*. Some local jurisdictions may adopt some of these standards and use them in inspections. In other cases the additional standards may be part of specifications that impose them as contract requirements.

The *NEC* is a bit more specific with regard to protecting cables from damage than it is with regard to support. In addition to the general statement about protecting cables from damage by the normal use of a building, Sections 800.6, 820.6, and 830.7 all refer to Section 300.4(D), which covers cables run parallel to framing members. The other requirements for cable protection in Section 300.4 are not referenced except for NPBC cables. These installations must comply with all the protection requirements of Section 300.4 according to Section 830.56. From a practical standpoint, in order for cables to be functional and remain functional, they must be protected from all types of potential damage, especially damage in concealed spaces. Therefore, as a practical matter, many installers may want to apply all of Section 300.4 to their cabling. Nevertheless, except for NPBC cables, there is no such reference in Chapter 8.

SUMMARY

Chapter 8 of the *NEC* covers communications systems. Included in this category are communications circuits most commonly used for telephones, antennas used for radio or TV reception, community antenna (CATV) circuits such those used for cable or satellite TV, and network-powered broadband communications (NPBC) circuits that have multiple uses. All of these systems and circuits are considered to be power-limited. Since Chapter 8 is independent of the other chapters of the *NEC*, it does not modify the other rules of the *Code,* and the other rules apply only when referenced within Chapter 8. Articles 800, 810, 820, and 830 provide all the rules that are needed to safely install communications circuits of various types, with only a few references to the other chapters where needed.

In this chapter we examine the rules for three types of circuits—communications, CATV, and NPBC—and establish the fact that the rules for these types of circuits are very similar in most cases. We also look at the differences in requirements. In addition to the basic requirements that the three types have in common and the definitions shared by these types, we also examine the specific types of rules that apply to the circuits covered by Chapter 8. The issues covered include requirements for removing abandoned cables; for limiting the spread of fire and products of combustion; for cables outside and entering buildings; for protection and protectors; for grounding; and for wiring methods and cable types, including permitted uses of cables and requirements for installing cables and maintaining circuit separations.

Communications Circuits

I. General

800.1 Scope. This article covers telephone, telegraph (except radio), outside wiring for fire alarm and burglar alarm, and similar central station systems; and telephone systems not connected to a central station system but using similar types of equipment, methods of installation, and maintenance.

> FPN No. 1: For further information for fire alarm, guard tour, sprinkler waterflow, and sprinkler supervisory systems, see Article 760.

> FPN No. 2: For installation requirements of optical fiber cables, see Article 770.

> FPN No. 3: For installation requirements for network-powered broadband communications circuits, see Article 830.

800.2 Definitions. See Article 100. For purposes of this article, the following additional definitions apply.

Abandoned Communications Cable. Installed communications cable that is not terminated at both ends at a connector or other equipment and not identified for future use with a tag.

Block. A square or portion of a city, town, or village enclosed by streets and including the alleys so enclosed, but not any street.

Cable. A factory assembly of two or more conductors having an overall covering.

Cable Sheath. A covering over the conductor assembly that may include one or more metallic members, strength members, or jackets.

Exposed. A circuit that is in such a position that, in case of failure of supports and insulation, contact with another circuit may result.

> FPN: See Article 100 for two other definitions of *Exposed*.

Point of Entrance. Within a building, the point at which the wire or cable emerges from an external wall, from a concrete floor slab, or from a rigid

---

Source: NFPA 70, *National Electrical Code®*, NFPA, Quincy, MA, 2002 edition.

metal conduit or an intermediate metal conduit grounded to an electrode in accordance with 800.40(B).

Premises. The land and buildings of a user located on the user side of the utility-user network point of demarcation.

Wire. A factory assembly of one or more insulated conductors without an overall covering.

800.3 Hybrid Power and Communications Cables. The provisions of 780.6 shall apply for listed hybrid power and communications cables in closed-loop and programmed power distribution.

> FPN: See 800.51(I) for hybrid power and communications cable in other applications.

800.4 Equipment. Equipment intended to be electrically connected to a telecommunications network shall be listed for the purpose. Installation of equipment shall also comply with 110.3(B).

> FPN: One way to determine applicable requirements is to refer to UL 1950-1993, *Standard for Safety of Information Technology Equipment, Including Electrical Business Equipment,* third edition; UL 1459-1995, *Standard for Safety, Telephone Equipment,* third edition; or UL 1863-1995, *Standard for Safety, Communications Circuit Accessories,* second edition. For information on listing requirements for communications raceways, see UL 2024-1995, *Standard for Optical Fiber Raceways.*

Exception: This listing requirement shall not apply to test equipment that is intended for temporary connection to a telecommunications network by qualified persons during the course of installation, maintenance, or repair of telecommunications equipment or systems.

800.5 Access to Electrical Equipment Behind Panels Designed to Allow Access. Access to electrical equipment shall not be denied by an accumulation of wires and cables that prevents removal of panels, including suspended ceiling panels.

800.6 Mechanical Execution of Work. Communications circuits and equipment shall be installed in a neat and workmanlike manner. Cables installed exposed on the outer surface of ceiling and sidewalls shall be supported by the structural components of the building structure in such a manner that the cable is not be damaged by normal building use. Such cables shall be attached to structural components by straps, staples, hangers, or similar fittings designed and installed so as not to damage the cable. The installation shall also conform with 300.4(D).

800.8 Hazardous (Classified) Locations. Communications circuits and equipment installed in a location that is classified in accordance with Article 500 shall comply with the applicable requirements of Chapter 5.

II. Conductors Outside and Entering Buildings

800.10 Overhead Communications Wires and Cables. Overhead communications wires and cables entering buildings shall comply with 800.10(A) and (B).

(A) On Poles and In-Span. Where communications wires and cables and electric light or power conductors are supported by the same pole or run parallel to each other in-span, the conditions described in 800.10(A)(1) through (A)(4) shall be met.

(1) Relative Location. Where practicable, the communications wires and cables shall be located below the electric light or power conductors.

(2) Attachment to Crossarms. Communications wires and cables shall not be attached to a cross-arm that carries electric light or power conductors.

(3) Climbing Space. The climbing space through communications wires and cables shall comply with the requirements of 225.14(D).

(4) Clearance. Supply service drops of 0–750 volts running above and parallel to communications service drops shall have a minimum separation of 300 mm (12 in.) at any point in the span, including the point of and at their attachment to the building, provided the nongrounded conductors are insulated and that a clearance of not less than 1.0 m (40 in.) is maintained between the two services at the pole.

(B) Above Roofs. Communications wires and cables shall have a vertical clearance of not less than 2.5 m (8 ft) from all points of roofs above which they pass.

Exception No. 1: Auxiliary buildings, such as garages and the like.

Exception No. 2: A reduction in clearance above only the overhanging portion of the roof to not less than 450 mm (18 in.) shall be permitted if (a) not more than 1.2 m (4 ft) of communications service-drop conductors pass above the roof overhang and (b) they are terminated at a through- or above-the-roof raceway or approved support.

Exception No. 3: Where the roof has a slope of not less than 100 mm in 300 mm (4 in. in 12 in.), a reduction in clearance to not less than 900 mm (3 ft) shall be permitted.

FPN: For additional information regarding overhead wires and cables, see ANSI C2-1997, *National Electric Safety Code*, Part 2 Safety Rules For Overhead Lines.

800.11 Underground Circuits Entering Buildings. Underground communications wires and cables entering buildings shall comply with 800.11(A) and (B).

(A) With Electric Light or Power Conductors. Underground communications wires and cables in a raceway, handhole, or manhole containing electric light, power, Class 1, or non–power-limited fire alarm circuit conductors shall be in a section separated from such conductors by means of brick, concrete, or tile partitions or by means of a suitable barrier.

(B) Underground Block Distribution. Where the entire street circuit is run underground and the circuit within the block is placed so as to be free from the likelihood of accidental contact with electric light or power circuits of over 300 volts to ground, the insulation requirements of 800.12(A) and (C) shall not apply, insulating supports shall not be required for the conductors, and bushings shall not be required where the conductors enter the building.

800.12 Circuits Requiring Primary Protectors. Circuits that require primary protectors as provided in 800.30 shall comply with 800.12(A), (B), and (C).

(A) Insulation, Wires, and Cables. Communications wires and cables without a metallic shield, running from the last outdoor support to the primary protector, shall be listed as being suitable for the purpose and shall have current-carrying capacity as specified in 800.30(A)(1)(b) or 800.30(A)(1)(c).

(B) On Buildings. Communications wires and cables in accordance with 800.12(A) shall be separated at least 100 mm (4 in.) from electric light or power conductors not in a raceway or cable or be permanently separated from conductors of the other system by a continuous and firmly fixed nonconductor in addition to the insulation on the wires, such as porcelain tubes or flexible tubing. Communications wires and cables in accordance with 800.12(A) exposed to accidental contact with electric light and power conductors operating at over 300 volts to ground and attached to buildings shall be separated from woodwork by being supported on glass, porcelain, or other insulating material.

Exception: Separation from woodwork shall not be required where fuses are omitted as provided for in 800.30(A)(1), or where conductors are used to extend circuits to a building from a cable having a grounded metal sheath.

(C) Entering Buildings. Where a primary protector is installed inside the building, the communications wires and cables shall enter the building either through a noncombustible, nonabsorbent insulating bushing or through a metal raceway. The insulating bushing shall not be required where the entering communications wires and cables (1) are in metal-sheathed cable, (2) pass through masonry, (3) meet the requirements of 800.12(A) and fuses are omitted as provided in 800.30(A)(1), or (4) meet the requirements of 800.12(A) and are used to extend circuits to a building from a cable having a grounded metallic sheath. Raceways or bushings shall slope upward from the outside or, where this cannot be done, drip loops shall be formed in the communications wires and cables immediately before they enter the building.

Raceways shall be equipped with an approved service head. More than one communications wire and cable shall be permitted to enter through a single raceway or bushing. Conduits or other metal raceways located ahead of the primary protector shall be grounded.

800.13 Lightning Conductors. Where practicable, a separation of at least 1.8 m (6 ft) shall be maintained between communications wires and cables on buildings and lightning conductors.

III. Protection

800.30 Protective Devices.

(A) Application. A listed primary protector shall be provided on each circuit run partly or entirely in aerial wire or aerial cable not confined within a block. Also, a listed primary protector shall be provided on each circuit, aerial or underground, located within the block containing the building served so as to be exposed to accidental contact with electric light or power conductors operating at over 300 volts to ground. In addition, where there exists a lightning exposure, each interbuilding circuit on a premises shall be protected by a listed primary protector at each end of the interbuilding circuit. Installation of primary protectors shall also comply with 110.3(B).

> FPN No. 1: On a circuit not exposed to accidental contact with power conductors, providing a listed primary protector in accordance with this article helps protect against other hazards, such as lightning and above-normal voltages induced by fault currents on power circuits in proximity to the communications circuit.

> FPN No. 2: Interbuilding circuits are considered to have a lightning exposure unless one or more of the following conditions exist:

> (1) Circuits in large metropolitan areas where buildings are close together and sufficiently high to intercept lightning.

(2) Interbuilding cable runs of 42 m (140 ft) or less, directly buried or in underground conduit, where a continuous metallic cable shield or a continuous metallic conduit containing the cable is bonded to each building grounding electrode system.

(3) Areas having an average of five or fewer thunderstorm days per year and earth resistivity of less than 100 ohm-meters. Such areas are found along the Pacific coast.

(1) Fuseless Primary Protectors. Fuseless-type primary protectors shall be permitted under any of the conditions given in (a) through (e).

(a) Where conductors enter a building through a cable with grounded metallic sheath member(s) and if the conductors in the cable safely fuse on all currents greater than the current-carrying capacity of the primary protector and of the primary protector grounding conductor

(b) Where insulated conductors in accordance with 800.12(A) are used to extend circuits to a building from a cable with an effectively grounded metallic sheath member(s) and if the conductors in the cable or cable stub, or the connections between the insulated conductors and the exposed plant, safely fuse on all currents greater than the current-carrying capacity of the primary protector, or the associated insulated conductors and of the primary protector grounding conductor

(c) Where insulated conductors in accordance with 800.12(A) or (B) are used to extend circuits to a building from other than a cable with a metallic sheath member(s) if (1) the primary protector is listed for this purpose, and (2) the connections of the insulated conductors to the exposed plant or the conductors of the exposed plant safely fuse on all currents greater than the current-carrying capacity of the primary protector, or the associated insulated conductors and of the primary protector grounding conductor

(d) Where insulated conductors in accordance with 800.12(A) are used to extend circuits aerially to a building from an unexposed buried or underground circuit

(e) Where insulated conductors in accordance with 800.12(A) are used to extend circuits to a building from cable with an effectively grounded metallic sheath member(s) and if (1) the combination of the primary protector and insulated conductors is listed for this purpose, and (2) the insulated conductors safely fuse on all currents greater than the current-carrying capacity of the primary protector and of the primary protector grounding conductor

(2) Fused Primary Protectors. Where the requirements listed under 800.30(A)(1)(a) through (1)(e) are not met, fused-type primary protectors shall be used. Fused-type primary protectors shall consist of an arrester connected between each line conductor and ground, a fuse in series with each line conductor, and an appropriate mounting arrangement. Primary protector terminals shall be marked to indicate line, instrument, and ground, as applicable.

(B) Location. The primary protector shall be located in, on, or immediately adjacent to the structure or building served and as close as practicable to the point of entrance.

> FPN: See 800.2 for the definition of *point of entrance.*

For purposes of this section, primary protectors located at mobile home service equipment located in sight from and not more than 9.0 m (30 ft) from the exterior wall of the mobile home it serves, or at a mobile home disconnecting means grounded in accordance with 250.32 and located in sight from and not more than 9.0 m (30 ft) from the exterior wall of the mobile home it serves, shall be considered to meet the requirements of this section.

> FPN: Selecting a primary protector location to achieve the shortest practicable primary protector grounding conductor helps limit potential differences between communications circuits and other metallic systems.

(C) Hazardous (Classified) Locations. The primary protector shall not be located in any hazardous (classified) location as defined in Article 500 or in the vicinity of easily ignitible material.

Exception: As permitted in 501.14, 502.14, and 503.12.

800.31 Primary Protector Requirements. The primary protector shall consist of an arrester connected between each line conductor and ground in an appropriate mounting. Primary protector terminals shall be marked to indicate line and ground as applicable.

> FPN: One way to determine applicable requirements for a listed primary protector is to refer to ANSI/UL 497-1995, *Standard for Protectors for Paired Conductor Communications Circuits.*

800.32 Secondary Protector Requirements. Where a secondary protector is installed in series with the indoor communications wire and cable between the primary protector and the equipment, it shall be listed for the purpose. The secondary protector shall provide means to safely limit currents to less than the current-carrying capacity of listed indoor communications wire and cable, listed telephone set line cords, and listed communications terminal equipment having ports for external wire line communications circuits. Any overvoltage protection, arresters, or grounding connection shall be connected on the equipment terminals side of the secondary protector current-limiting means.

FPN No. 1: One way to determine applicable requirements for a listed secondary protector is to refer to UL 497A-1996, *Standard for Secondary Protectors for Communications Circuits.*

FPN No. 2: Secondary protectors on exposed circuits are not intended for use without primary protectors.

800.33 Cable Grounding. The metallic sheath of communications cables entering buildings shall be grounded as close as practicable to the point of entrance or shall be interrupted as close to the point of entrance as practicable by an insulating joint or equivalent device.

FPN: See 800.2 for the definition of *point of entrance.*

IV. Grounding Methods

800.40 Cable and Primary Protector Grounding. The metallic member(s) of the cable sheath, where required to be grounded by 800.33, and primary protectors shall be grounded as specified in 800.40(A) through (D).

(A) Grounding Conductor.

(1) Insulation. The grounding conductor shall be insulated and shall be listed as suitable for the purpose.

(2) Material. The grounding conductor shall be copper or other corrosion-resistant conductive material, stranded or solid.

(3) Size. The grounding conductor shall not be smaller than 14 AWG.

(4) Length. The primary protector grounding conductor shall be as short as practicable. In one- and two-family dwellings, the primary protector grounding conductor shall be as short as practicable, not to exceed 6.0 m (20 ft) in length.

Exception: In one- and two-family dwellings where it is not practicable to achieve an overall maximum primary protector grounding conductor length of 6.0 m (20 ft), a separate communications ground rod meeting the minimum dimensional criteria of 800.40(B)(2)(2) shall be driven, the primary protector shall be grounded to the communications ground rod in accordance with 800.40(C), and the communications ground rod bonded to the power grounding electrode system in accordance with 800.40(D).

(5) Run in Straight Line. The grounding conductor shall be run to the grounding electrode in as straight a line as practicable.

(6) Physical Damage. Where necessary, the grounding conductor shall be guarded from physical damage. Where the grounding conductor is run in a metal raceway, both ends of the raceway shall be bonded to the grounding conductor or the same terminal or electrode to which the grounding conductor is connected.

(B) Electrode. The grounding conductor shall be connected in accordance with 800.40(B)(1) and (B)(2).

(1) In Buildings or Structures with Grounding Means. To the nearest accessible location on the following:

(1) The building or structure grounding electrode system as covered in 250.50
(2) The grounded interior metal water piping system, within 1.5 m (5 ft) from its point of entrance to the building, as covered in 250.52
(3) The power service accessible means external to enclosures as covered in 250.94
(4) The metallic power service raceway
(5) The service equipment enclosure
(6) The grounding electrode conductor or the grounding electrode conductor metal enclosure
(7) The grounding conductor or the grounding electrode of a building or structure disconnecting means that is grounded to an electrode as covered in 250.32.

For purposes of this section, the mobile home service equipment or the mobile home disconnecting means, as described in 800.30(B), shall be considered accessible.

(2) In Buildings or Structures Without Grounding Means. If the building or structure served has no grounding means, as described in 800.40(B)(1):

(1) To any one of the individual electrodes described in 250.52(A)(1), (2), (3), (4); or
(2) If the building or structure served has no grounding means, as described in 800.40(B)(1) or (B)(2)(1), to an effectively grounded metal structure or to a ground rod or pipe not less than 1.5 m (5 ft) in length and 12.7 mm (½ in.) in diameter, driven, where practicable, into permanently damp earth and separated from lightning conductors as covered in 800.13 and at least 1.8 m (6 ft) from electrodes of other systems. Steam or hot water pipes or air terminal conductors (lightning-rod conductors) shall not be employed as electrodes for protectors.

(C) Electrode Connection. Connections to grounding electrodes shall comply with 250.70. Connectors, clamps, fittings, or lugs used to attach grounding conductors and bonding jumpers to grounding electrodes or to each other that are to be concrete-encased or buried in the earth shall be suitable for its application.

(D) Bonding of Electrodes. A bonding jumper not smaller than 6 AWG copper or equivalent shall be connected between the communications grounding electrode and power grounding electrode system at the building or structure served where separate electrodes are used. Bonding together of all separate electrodes shall be permitted.

Exception: At mobile homes as covered in 800.41.

FPN No. 1: See 250.60 for use of air terminals (lightning rods).

FPN No. 2: Bonding together of all separate electrodes limits potential differences between them and between their associated wiring systems.

800.41 Primary Protector Grounding and Bonding at Mobile Homes.

(A) Grounding. Where there is no mobile home service equipment located in sight from, and not more than 9.0 m (30 ft) from, the exterior wall of the mobile home it serves, or there is no mobile home disconnecting means grounded in accordance with 250.32 and located within sight from, and not more than 9.0 m (30 ft) from, the exterior wall of the mobile home it serves, the primary protector ground shall be in accordance with 800.40(B)(2).

(B) Bonding. The primary protector grounding terminal or grounding electrode shall be bonded to the metal frame or available grounding terminal of the mobile home with a copper grounding conductor not smaller than 12 AWG under any of the following conditions:

(1) Where there is no mobile home service equipment or disconnecting means as in 800.41(A)
(2) Where the mobile home is supplied by cord and plug

V. Communications Wires and Cables Within Buildings

800.48 Raceways for Communications Wires and Cables. Where communications wire and cables are installed in a raceway, the raceway shall be either of a type permitted in Chapter 3 and installed in accordance with Chapter 3 or a listed nonmetallic raceway complying with 800.51(J), (K), or (L), as applicable, and installed in accordance with 362.24 through 362.56, where the requirements applicable to electrical nonmetallic tubing apply.

Exception: Conduit fill restrictions shall not apply.

800.49 Fire Resistance of Communications Wires and Cables. Communications wires and cables installed as wiring within a building shall be listed as being resistant to the spread of fire in accordance with 800.50 and 800.51.

800.50 Listing, Marking, and Installation of Communications Wires and Cables. Communications wires and cables installed as wiring within buildings shall be listed as being suitable for the purpose and installed in accordance with 800.52. Communications cables and undercarpet communications wires shall be marked in accordance with Table 800.50. The cable voltage rating shall not be marked on the cable or on the undercarpet communications wire.

FPN: Voltage markings on cables may be misinterpreted to suggest that the cables may be suitable for Class 1, electric light, and power applications.

Exception No. 1: Voltage markings shall be permitted where the cable has multiple listings and voltage marking is required for one or more of the listings.

Exception No. 2: Listing and marking shall not be required where the cable enters the building from the outside and is continuously enclosed in a rigid metal conduit system or an intermediate metal conduit system and such conduit systems are grounded to an electrode in accordance with 800.40(B).

Exception No. 3: Listing and marking shall not be required where the length of the cable within the building, measured from its point of entrance, does not exceed 15 m (50 ft) and the cable enters the building from the outside and is terminated in an enclosure or on a listed primary protector.

FPN No. 1: Splice cases or terminal boxes, both metallic and plastic types, are typically used as enclosures for splicing or terminating telephone cables.

FPN No. 2: This exception limits the length of unlisted outside plant cable to 15 m (50 ft), while 800.30(B) requires that the primary protector be located as close as practicable to the point at which the cable enters the building. Therefore, in installations requiring a primary protector, the outside plant cable may not be permitted to extend 15 m (50 ft) into the building if it is practicable to place the primary protector closer than 15 m (50 ft) to the entrance point.

Exception No. 4: Multipurpose cables shall be considered as being suitable for the purpose and shall be permitted to substitute for communications cables as provided for in 800.53(G).

FPN No. 1: Cable types are listed in descending order of fire resistance rating, and multipurpose cables are listed above communications cables because multipurpose cables may substitute for communications cables.

FPN No. 2: See the referenced sections for permitted uses.

Table 800.50   Cable Markings

Cable Marking	Type	Reference
MPP	Multipurpose plenum cable	800.51(G) and 800.53(A)
CMP	Communications plenum cable	800.51(A) and 800.53(A)
MPR	Multipurpose riser cable	800.51(G) and 800.53(B)
CMR	Communications riser cable	800.51(B) and 800.53(B)
MPG	Multipurpose general-purpose cable	800.51(G) and 800.53(D) and (E)(1)
CMG	Communications general-purpose cable	800.51(C) and 800.53(D) and (E)(1)
MP	Multipurpose general-purpose cable	800.51(G) and 800.53(D) and (E)(1)
CM	Communications general-purpose cable	800.51(D) and 800.53(D) and (E)(1)
CMX	Communications cable, limited use	800.51(E) and 800.53(C), (D), and (E)
CMUC	Undercarpet communications wire and cable	800.51(F) and 800.53(F)(6)

800.51 Listing Requirements for Communications Wires and Cables and Communications Raceways. Communications wires and cables shall have a voltage rating of not less than 300 volts and shall be listed in accordance with 800.51(A) through (J), and communications raceways shall be listed in accordance with 800.51(K) through (L). Conductors in communications cables, other than in a coaxial cable, shall be copper.

FPN: See 800.4 for listing requirement for equipment.

(A) Type CMP. Type CMP communications plenum cable shall be listed as being suitable for use in ducts, plenums, and other spaces used for environmental air and shall also be listed as having adequate fire-resistant and low smoke-producing characteristics.

> FPN: One method of defining low smoke-producing cables is by establishing an acceptable value of the smoke produced when tested in accordance with NFPA 262-1999, *Standard Method of Test for Flame Travel and Smoke of Wire and Cables for Use in Air-Handling Spaces*, to a maximum peak optical density of 0.5 and a maximum average optical density of 0.15. Similarly, one method of defining fire-resistant cables is by establishing a maximum allowable flame travel distance of 1.52 m (5 ft) when tested in accordance with the same test.

(B) Type CMR. Type CMR communications riser cable shall be listed as being suitable for use in a vertical run in a shaft or from floor to floor and shall also be listed as having fire-resistant characteristics capable of preventing the carrying of fire from floor to floor.

> FPN: One method of defining fire-resistant characteristics capable of preventing the carrying of fire from floor to floor is that the cables pass the requirements of ANSI/UL 1666-1997, *Standard Test for Flame Propagation Height of Electrical and Optical-Fiber Cable Installed Vertically in Shafts*.

(C) Type CMG. Type CMG general-purpose communications cable shall be listed as being suitable for general-purpose communications use, with the exception of risers and plenums, and shall also be listed as being resistant to the spread of fire.

> FPN: One method of defining *resistant to the spread of fire* is for the damage (char length) not to exceed 1.5 m (4 ft 11 in.) when performing the vertical flame test for cables in cable trays, as described in CSA C22.2 No. 0.3-M 1985, *Test Methods for Electrical Wires and Cables*.

(D) Type CM. Type CM communications cable shall be listed as being suitable for general-purpose communications use, with the exception of risers and plenums, and shall also be listed as being resistant to the spread of fire.

> FPN: One method of defining *resistant to the spread of fire* is that the cables do not spread fire to the top of the tray in the vertical-tray flame test in ANSI/UL 1581-1991, *Reference Standard for Electrical Wires, Cables and Flexible Cords*. Another method of defining *resistant to the spread of fire* is for the damage (char length) not to exceed 1.5 m (4 ft 11 in.) when performing the vertical flame test for cables in cable trays, as

described in CSA C22.2 No. 0.3-M-1985, *Test Methods for Electrical Wires and Cables.*

(E) Type CMX. Type CMX limited-use communications cable shall be listed as being suitable for use in dwellings and for use in raceway and shall also be listed as being resistant to flame spread.

FPN: One method of determining that cable is resistant to flame spread is by testing the cable to the VW-1 (vertical-wire) flame test in ANSI/UL 1581-1991, *Reference Standard for Electrical Wires, Cables and Flexible Cords.*

(F) Type CMUC Undercarpet Wire and Cable. Type CMUC undercarpet communications wire and cable shall be listed as being suitable for undercarpet use and shall also be listed as being resistant to flame spread.

FPN: One method of determining that cable is resistant to flame spread is by testing the cable to the VW-1 (vertical-wire) flame test in ANSI/UL 1581-1991, *Reference Standard for Electrical Wires, Cables and Flexible Cords.*

(G) Multipurpose (MP) Cables. Until July 1, 2003, cables that meet the requirements for Types CMP, CMR, CMG, and CM and also satisfy the requirements of 760.71(B) for multiconductor cables and 760.71(H) for coaxial cables shall be permitted to be listed and marked as multipurpose cable Types MPP, MPR, MPG, and MP, respectively.

(H) Communications Wires. Communications wires, such as distributing frame wire and jumper wire, shall be listed as being resistant to the spread of fire.

FPN: One method of defining *resistant to the spread of fire* is that the cables do not spread fire to the top of the tray in the vertical-tray flame test in ANSI/UL 1581-1991, *Reference Standard for Electrical Wires, Cables and Flexible Cords.* Another method of defining *resistant to the spread of fire* is for the damage (char length) not to exceed 1.5 m (4 ft 11 in.) when performing the vertical flame test for cables in cable trays, as described in CSA C22.2 No. 0.3-M-1985, *Test Methods for Electrical Wires and Cables.*

(I) Hybrid Power and Communications Cable. Listed hybrid power and communications cable shall be permitted where the power cable is a listed Type NM or NM-B conforming to the provisions of Article 334, and the communications cable is a listed Type CM, the jackets on the listed NM or NM-B and listed CM cables are rated for 600 volts minimum, and the hybrid cable is listed as being resistant to the spread of fire.

FPN: One method of defining *resistant to the spread of fire* is that the cables do not spread fire to the top of the tray in the vertical-tray flame test in ANSI/UL 1581-1991, *Reference Standard for Electrical Wires, Cables and Flexible Cords*. Another method of defining *resistant to the spread of fire* is for the damage (char length) not to exceed 1.5 m (4 ft 11 in.) when performing the vertical flame test for cables in cable trays, as described in CSA C22.2 No. 0.3-M-1985, *Test Methods for Electrical Wires and Cables*.

(J) Plenum Communications Raceways. Plenum communications raceways listed as plenum optical fiber raceways shall be permitted for use in ducts, plenums, and other spaces used for environmental air and shall also be listed as having adequate fire-resistant and low smoke-producing characteristics.

(K) Riser Communications Raceway. Riser communications raceways shall be listed as having adequate fire-resistant characteristics capable of preventing the carrying of fire from floor to floor.

(L) General-Purpose Communications Raceway. General-purpose communications raceways shall be listed as being resistant to the spread of fire.

800.52 Installation of Communications Wires, Cables, and Equipment. Communications wires and cables from the protector to the equipment or, where no protector is required, communications wires and cables attached to the outside or inside of the building shall comply with 800.52(A) through (E).

(A) Separation from Other Conductors.

(1) In Raceways, Boxes, and Cables.

(a) Other Power-Limited Circuits. Communications cables shall be permitted in the same raceway or enclosure with cables of any of the following:

(1) Class 2 and Class 3 remote-control, signaling, and power-limited circuits in compliance with Article 725
(2) Power-limited fire alarm systems in compliance with Article 760
(3) Nonconductive and conductive optical fiber cables in compliance with Article 770
(4) Community antenna television and radio distribution systems in compliance with Article 820
(5) Low-power network-powered broadband communications circuits in compliance with Article 830

(b) Class 2 and Class 3 Circuits. Class 1 circuits shall not be run in the same cable with communications circuits. Class 2 and Class 3 circuit conductors shall be permitted in the same cable with communications circuits, in which case the Class 2 and Class 3 circuits shall be classified as communications circuits and shall meet the requirements of this article. The cables shall be listed as communications cables or multipurpose cables.

Exception: Cables constructed of individually listed Class 2, Class 3, and communications cables under a common jacket shall not be required to be classified as communications cable. The fire-resistance rating of the composite cable shall be determined by the performance of the composite cable.

(c) Electric Light, Power, Class 1, Non–Power-Limited Fire Alarm, and Medium Power Network-Powered Broadband Communications Circuits in Raceways, Compartments, and Boxes. Communications conductors shall not be placed in any raceway, compartment, outlet box, junction box, or similar fitting with conductors of electric light, power, Class 1, non–power-limited fire alarm or medium power network-powered broadband communications circuits.

Exception No. 1: Where all of the conductors of electric light, power, Class 1, non–power-limited fire alarm, and medium power network-powered broadband communications circuits are separated from all of the conductors of communications circuits by a barrier.

Exception No. 2: Power conductors in outlet boxes, junction boxes, or similar fittings or compartments where such conductors are introduced solely for power supply to communications equipment. The power circuit conductors shall be routed within the enclosure to maintain a minimum of 6 mm (0.25 in.) separation from the communications circuit conductors.

Exception No. 3: As permitted by 620.36.

(2) Other Applications. Communications wires and cables shall be separated at least 50 mm (2 in.) from conductors of any electric light, power, Class 1, non–power-limited fire alarm, or medium power network-powered broadband communications circuits.

Exception No. 1: Where either (1) all of the conductors of the electric light, power, Class 1, non–power-limited fire alarm, and medium power network-powered broadband communications circuits are in a raceway or in metal-sheathed, metal-clad, nonmetallic-sheathed, Type AC, or Type UF cables, or (2) all of the conductors of communications circuits are encased in raceway.

Exception No. 2: Where the communications wires and cables are permanently separated from the conductors of electric light, power, Class 1, non–power-limited fire

*alarm, and medium power network-powered broadband communications circuits
by a continuous and firmly fixed nonconductor, such as porcelain tubes or flexible tubing, in addition to the insulation on the wire.*

(B) Spread of Fire or Products of Combustion. Installations in hollow
spaces, vertical shafts, and ventilation or air-handling ducts shall be made
so that the possible spread of fire or products of combustion is not substantially increased. Openings around penetrations through fire resistance-rated
walls, partitions, floors, or ceilings shall be firestopped using approved
methods to maintain the fire resistance rating.

The accessible portion of abandoned communications cables shall not
be permitted to remain.

> FPN: Directories of electrical construction materials published by
> qualified testing laboratories contain many listing installation restric-
> tions necessary to maintain the fire-resistive rating of assemblies
> where penetrations or openings are made.

(C) Equipment in Other Space Used for Environmental Air. Section
300.22(C) shall apply.

(D) Cable Trays. Types MPP, MPR, MPG, and MP multipurpose cables
and Types CMP, CMR, CMG, and CM communications cables shall be permitted to be installed in cable trays. Communications raceways, as described in 800.51, shall be permitted to be installed in cable trays.

(E) Support of Conductors. Raceways shall be used for their intended
purpose. Communications cables or wires shall not be strapped, taped, or
attached by any means to the exterior of any conduit or raceway as a means
of support.

*Exception: Overhead (aerial) spans of communications cables or wires shall be per-
mitted to be attached to the exterior of a raceway-type mast intended for the attachment and support of such conductors.*

**800.53 Applications of Listed Communications Wires and Cables and
Communications Raceways.** Communications wires and cables shall comply with the requirements of 800.53(A) through (F) or where cable substitutions are made in accordance with 800.53(G).

(A) Plenum. Cables installed in ducts, plenums, and other spaces used for
environmental air shall be Type CMP. Abandoned cables shall not be permitted to remain. Types CMP, CMR, CMG, CM, and CMX and communications wire installed in compliance with 300.22 shall be permitted. Listed
plenum communications raceways shall be permitted to be installed in

ducts and plenums as described in 300.22(B) and in other spaces used for environmental air as described in 300.22(C). Only Type CMP cable shall be permitted to be installed in these raceways.

(B) Riser. Cables installed in risers shall comply with 800.53(B)(1), (B)(2), or (B)(3).

(1) Cables in Vertical Runs. Cables installed in vertical runs and penetrating more than one floor, or cables installed in vertical runs in a shaft, shall be Type CMR. Floor penetrations requiring Type CMR shall contain only cables suitable for riser or plenum use. Abandoned cables shall not be permitted to remain. Listed riser communications raceways shall be permitted to be installed in vertical riser runs in a shaft from floor to floor. Only Type CMR and CMP cables shall be permitted to be installed in these raceways.

(2) Metal Raceways or Fireproof Shafts. Listed communications cables shall be encased in a metal raceway or located in a fireproof shaft having firestops at each floor.

(3) One- and Two-Family Dwellings. Type CM and CMX cable shall be permitted in one- and two-family dwellings.

FPN: See 800.52(B) for firestop requirements for floor penetrations.

(C) Distributing Frames and Cross-Connect Arrays. Listed communications wire and Types CMP, CMR, CMG, and CM communications cables shall be used in distributing frames and cross-connect arrays.

(D) Cable Trays. Types MPP, MPR, MPG, and MP multipurpose cables and Types CMP, CMR, CMG, and CM communications cables shall be permitted to be installed in cable trays.

(E) Other Wiring Within Buildings. Cables installed in building locations other than the locations covered in 800.53(A) through (D) shall be in accordance with 800.53(E)(1) through (E)(6).

(1) General. Cables shall be Type CMG or Type CM. Listed communications general-purpose raceways shall be permitted. Only Types CMG, CM, CMR, or CMP cables shall be permitted to be installed in general-purpose communications raceways.

(2) In Raceways. Listed communications wires that are enclosed in a raceway of a type included in Chapter 3 shall be permitted.

(3) Nonconcealed Spaces. Type CMX communications cable shall be permitted to be installed in nonconcealed spaces where the exposed length of cable does not exceed 3 m (10 ft).

(4) One- and Two-Family Dwellings. Type CMX communications cable less than 6 mm (0.25 in.) in diameter shall be permitted to be installed in one- and two-family dwellings.

(5) Multi-Family Dwellings. Type CMX communications cable less than 6 mm (0.25 in.) in diameter shall be permitted to be installed in nonconcealed spaces in multi-family dwellings.

(6) Under Carpets. Type CMUC undercarpet communications wires and cables shall be permitted to be installed under carpet.

(F) Hybrid Power and Communications Cable. Hybrid power and communications cable listed in accordance with 800.51(J) shall be permitted to be installed in one- and two-family dwellings.

(G) Cable Substitutions. The uses and permitted substitutions for communications cables listed in Table 800.53 shall be considered suitable for the purpose and shall be permitted.

FPN: For information on Types CMP, CMR, CMG, CM, and CMX cables, see 800.51.

Table 800.53 Cable Uses and Permitted Substitutions

Cable Type	Use	References	Permitted Substitutions
CMP	Communications plenum cable	800.53(A)	MPP
CMR	Communications riser cable	800.53(B)	MPP, CMP, MPR
CMG, CM	Communications general-purpose cable	800.53(E)(1)	MPP, CMP, MPR, CMR, MPG, MP
CMX	Communications cable, limited use	800.53(E)	MPP, CMP, MPR, CMR, MPG, MP, CMG, CM

Note: See Figure 800.53, Cable substitution hierarchy.

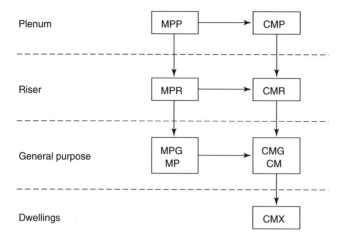

Plenum

Riser

General purpose

Dwellings

Type CM—Communications cables

Type MP—Multipurpose cable

A → B   Cable A shall be permitted to be used in place of cable B.

Figure 800.53 Cable substitution hierarchy.

Community Antenna Television and Radio Distribution Systems

I. General

820.1 Scope. This article covers coaxial cable distribution of radio frequency signals typically employed in community antenna television (CATV) systems.

820.2 Definitions. See Article 100. For the purposes of this article, the following additional definitions apply.

Abandoned Coaxial Cable. Installed coaxial cable that is not terminated at equipment other than a coaxial connector and not identified for future use with a tag.

Exposed. An exposed cable is one that is in such a position that, in case of failure of supports and insulation, contact with another circuit could result.

    FPN: See Article 100 for two other definitions of *exposed*.

Point of Entrance. The point within a building at which the cable emerges from an external wall, from a concrete floor slab, or from a rigid metal conduit or an intermediate metal conduit grounded to an electrode in accordance with 820.40(B).

Premises. The land and buildings of a user located on the user side of utility-user network point of demarcation.

820.3 Locations and Other Articles. Circuits and equipment shall comply with 820.3(A) through (G).

(A) Spread of Fire or Products of Combustion. Section 300.21 shall apply. The accessible portion of abandoned coaxial cables shall not be permitted to remain.

(B) Ducts, Plenums, and Other Air-Handling Spaces. Section 300.22, where installed in ducts or plenums or other spaces used for environmental air, shall apply.

Source: NFPA 70, *National Electrical Code®*, NFPA, Quincy, MA, 2002 edition.

Exception: As permitted in 820.53(A).

(C) Installation and Use. Section 110.3 shall apply.

(D) Installations of Conductive and Nonconductive Optical Fiber Cables. Article 770 shall apply.

(E) Communications Circuits. Article 800 shall apply.

(F) Network-Powered Broadband Communications Systems. Article 830 shall apply.

(G) Alternate Wiring Methods. The wiring methods of Article 830 shall be permitted to substitute for the wiring methods of Article 820.

> FPN: Use of Article 830 wiring methods will facilitate the upgrading of Article 820 installations to network-powered broadband applications.

820.4 Energy Limitations. The coaxial cable shall be permitted to deliver low-energy power to equipment that is directly associated with the radio frequency distribution system if the voltage is not over 60 volts and if the current supply is from a transformer or other device that has energy-limiting characteristics.

820.5 Access to Electrical Equipment Behind Panels Designed to Allow Access. Access to electrical equipment shall not be denied by an accumulation of wires and cables that prevents removal of panels, including suspended ceiling panels.

820.6 Mechanical Execution of Work. Community antenna television and radio distribution systems shall be installed in a neat and workmanlike manner. Cables installed exposed on the surface of ceiling and sidewalls shall be supported by the structural components of the building structure in such a manner that the cable is not damaged by normal building use. Such cables shall be attached to structural components by straps, staples, hangers, or similar fittings designed and installed so as not to damage the cable. The installation shall also conform with 300.4(D).

II. Cables Outside and Entering Buildings

820.10 Outside Cables. Coaxial cables, prior to the point of grounding, as defined in 820.33, shall comply with 820.10(A) through (F).

(A) On Poles. Where practicable, conductors on poles shall be located below the electric light, power, Class 1, or non–power-limited fire alarm circuit conductors and shall not be attached to a crossarm that carries electric light or power conductors.

(B) Lead-in Clearance. Lead-in or aerial-drop cables from a pole or other support, including the point of initial attachment to a building or structure, shall be kept away from electric light, power, Class 1, or non–power-limited fire alarm circuit conductors so as to avoid the possibility of accidental contact.

Exception: Where proximity to electric light, power, Class 1, or non–power-limited fire alarm circuit service conductors cannot be avoided, the installation shall be such as to provide clearances of not less than 300 mm (12 in.) from light, power, Class 1, or non–power-limited fire alarm circuit service drops. The clearance requirement shall apply at all points along the drop, and it shall increase to 1.02 m (40 in.) at the pole.

(C) On Masts. Aerial cable shall be permitted to be attached to an above-the-roof raceway mast that does not enclose or support conductors of electric light or power circuits.

(D) Above Roofs. Cables shall have a vertical clearance of not less than 2.5 m (8 ft) from all points of roofs above which they pass.

Exception No. 1: Auxiliary buildings such as garages and the like.

Exception No. 2: A reduction in clearance above only the overhanging portion of the roof to not less than 450 mm (18 in.) shall be permitted if (1) not more than 1.2 m (4 ft) of communications service drop conductors pass above the roof overhang, and (2) they are terminated at a raceway mast or other approved support.

Exception No. 3: Where the roof has a slope of not less than 100 mm (4 in.) in 300 mm (12 in.), a reduction in clearance to not less than 900 mm (3 ft) shall be permitted.

(E) Between Buildings. Cables extending between buildings and also the supports or attachment fixtures shall be acceptable for the purpose and shall have sufficient strength to withstand the loads to which they may be subjected.

Exception: Where a cable does not have sufficient strength to be self-supporting, it shall be attached to a supporting messenger cable that, together with the attachment fixtures or supports, shall be acceptable for the purpose and shall have sufficient strength to withstand the loads to which they may be subjected.

(F) On Buildings. Where attached to buildings, cables shall be securely fastened in such a manner that they will be separated from other conductors in accordance with 820.10(F)(1), (F)(2), and (F)(3).

(1) Electric Light or Power. The coaxial cable shall have a separation of at least 100 mm (4 in.) from electric light, power, Class 1, or non–power-limited fire alarm circuit conductors not in raceway or cable or be permanently separated from conductors of the other system by a continuous and firmly fixed nonconductor in addition to the insulation on the wires.

(2) Other Communications Systems. Coaxial cable shall be installed so that there will be no unnecessary interference in the maintenance of the separate systems. In no case shall the conductors, cables, messenger strand, or equipment of one system cause abrasion to the conductors, cable, messenger strand, or equipment of any other system.

(3) Lightning Conductors. Where practicable, a separation of at least 1.8 m (6 ft) shall be maintained between any coaxial cable and lightning conductors.

> FPN: For additional information regarding overhead wires and cables, see ANSI C2-1997, *National Electric Safety Code*, Part 2, Safety Rules for Overhead Lines.

820.11 Entering Buildings.

(A) Underground Systems. Underground coaxial cables in a duct, pedestal, handhole, or manhole that contains electric light or power conductors or Class 1 circuits shall be in a section permanently separated from such conductors by means of a suitable barrier.

(B) Direct-Buried Cables and Raceways. Direct-buried coaxial cable shall be separated at least 300 mm (12 in.) from conductors of any light or power or Class 1 circuit.

Exception No. 1: Where electric service conductors or coaxial cables are installed in raceways or have metal cable armor.

Exception No. 2: Where electric light or power branch-circuit or feeder conductors or Class 1 circuit conductors are installed in a raceway or in metal-sheathed, metal-clad, or Type UF or Type USE cables; or the coaxial cables have metal cable armor or are installed in a raceway.

III. Protection

820.33 Grounding of Outer Conductive Shield of a Coaxial Cable. The outer conductive shield of the coaxial cable shall be grounded at the building premises as close to the point of cable entrance or attachment as practicable.

For purposes of this section, grounding located at mobile home service equipment located in sight from, and not more than 9.0 m (30 ft) from, the exterior wall of the mobile home it serves, or at a mobile home disconnecting means grounded in accordance with 250.32 and located in sight from and not more than 9.0 m (30 ft) from the exterior wall of the mobile home it serves, shall be considered to meet the requirements of this section.

> FPN: Selecting a grounding location to achieve the shortest practicable grounding conductor helps limit potential differences between CATV and other metallic systems.

(A) Shield Grounding. Where the outer conductive shield of a coaxial cable is grounded, no other protective devices shall be required.

(B) Shield Protection Devices. Grounding of a coaxial drop cable shield by means of a protective device that does not interrupt the grounding system within the premises shall be permitted.

IV. Grounding Methods

820.40 Cable Grounding. Where required by 820.33, the shield of the coaxial cable shall be grounded as specified in 820.40(A) through (D).

(A) Grounding Conductor.

(1) Insulation. The grounding conductor shall be insulated and shall be listed as suitable for the purpose.

(2) Material. The grounding conductor shall be copper or other corrosion-resistant conductive material, stranded or solid.

(3) Size. The grounding conductor shall not be smaller than 14 AWG. It shall have a current-carrying capacity approximately equal to that of the outer conductor of the coaxial cable. The grounding conductor shall not be required to exceed 6 AWG.

(4) Length. The grounding conductor shall be as short as practicable. In one- and two-family dwellings, the grounding conductor shall be as short as practicable, not to exceed 6.0 m (20 ft) in length.

Exception: In one- and two-family dwellings where it is not practicable to achieve an overall maximum grounding conductor length of 6.0 m (20 ft), a separate ground as specified in 250.52(A)(5), (6), or (7) shall be used, the grounding conductor shall be grounded to the separate ground in accordance with 250.70, and the separate ground bonded to the power grounding electrode system in accordance with 820.40(D).

(5) Run in Straight Line. The grounding conductor shall be run to the grounding electrode in as straight a line as practicable.

(6) Physical Protection. Where subject to physical damage, the grounding conductor shall be adequately protected. Where the grounding conductor is run in a metal raceway, both ends of the raceway shall be bonded to the grounding conductor or the same terminal or electrode to which the grounding conductor is connected.

(B) Electrode. The grounding conductor shall be connected in accordance with 820.40(B)(1) and (B)(2).

(1) In Buildings or Structures with Grounding Means. To the nearest accessible location on the following:

(1) The building or structure grounding electrode system as covered in 250.50;
(2) The grounded interior metal water piping system, within 1.52 m (5 ft) from its point of entrance to the building, as covered in 250.52;
(3) The power service accessible means external to enclosures as covered in 250.94;
(4) The metallic power service raceway;
(5) The service equipment enclosure;
(6) The grounding electrode conductor or the grounding electrode conductor metal enclosure; or
(7) The grounding conductor or the grounding electrode of a building or structure disconnecting means that is grounded to an electrode as covered in 250.32.

(2) In Buildings or Structures Without Grounding Means. If the building or structure served has no grounding means, as described in 820.40(B)(1):

(1) To any one of the individual electrodes described in 250.52(A)(1), (2), (3), (4); or,
(2) If the building or structure served has no grounding means, as described in 820.40(B)(1) or (B)(2)(1), to an effectively grounded metal structure or to any one of the individual electrodes described in 250.52(A)(5), (6), and (7).

(C) Electrode Connection. Connections to grounding electrodes shall comply with 250.70.

(D) Bonding of Electrodes. A bonding jumper not smaller than 6 AWG copper or equivalent shall be connected between the antenna systems grounding electrode and the power grounding electrode system at the building or structure served where separate electrodes are used.

Exception: At mobile homes as covered in 820.42.

FPN No. 1: See 250.60 for use of air terminals (lightning rods).

FPN No. 2: Bonding together of all separate electrodes limits potential differences between them and between their associated wiring systems.

820.41 Equipment Grounding. Unpowered equipment and enclosures or equipment powered by the coaxial cable shall be considered grounded where connected to the metallic cable shield.

820.42 Bonding and Grounding at Mobile Homes.

(A) Grounding. Where there is no mobile home service equipment located in sight from, and not more than 9.0 m (30 ft) from, the exterior wall of the mobile home it serves or there is no mobile home disconnecting means grounded in accordance with 250.32 and located within sight from, and not more than 9.0 m (30 ft) from, the exterior wall of the mobile home it serves, the coaxial cable shield ground, or surge arrester ground, shall be in accordance with 820.40(B)(2).

(B) Bonding. The coaxial cable shield grounding terminal, surge arrester grounding terminal, or grounding electrode shall be bonded to the metal frame or available grounding terminal of the mobile home with a copper grounding conductor not smaller than 12 AWG under any of the following conditions:

(1) Where there is no mobile home service equipment or disconnecting means as in 820.42(A)
(2) Where the mobile home is supplied by cord and plug

V. Cables Within Buildings

820.49 Fire Resistance of CATV Cables. Coaxial cables installed as wiring within buildings shall be listed as being resistant to the spread of fire in accordance with 820.50 and 820.51.

820.50 Listing, Marking, and Installation of Coaxial Cables. Coaxial cables in a building shall be listed as being suitable for the purpose, and cables shall be marked in accordance with Table 820.50. The cable voltage rating shall not be marked on the cable.

> FPN: Voltage markings on cables could be misinterpreted to suggest that the cables may be suitable for Class 1, electric light, and power applications.

Exception No. 1: Voltage markings shall be permitted where the cable has multiple listings and voltage marking is required for one or more of the listings.

Exception No. 2: Listing and marking shall not be required where the cable enters the building from the outside and is run in rigid metal conduit or intermediate metal conduit, and such conduits are grounded to an electrode in accordance with 820.40(B).

Table 820.50 Cable Markings

Cable Marking	Type	Reference
CATVP	CATV plenum cable	820.51(A) and 820.53(A)
CATVR	CATV riser cable	820.51(B) and 820.53(B)
CATV	CATV cable	820.51(C) and 820.53(C)
CATVX	CATV cable, limited use	820.51(D) and 820.53(C)

Exception No. 3: Listing and marking shall not be required where the length of the cable within the building, measured from its point of entrance, does not exceed 15 m (50 ft) and the cable enters the building from the outside and is terminated at a grounding block.

> FPN No. 1: Cable types are listed in descending order of fire-resistance rating.

> FPN No. 2: See the referenced sections for listing requirements and permitted uses.

820.51 Additional Listing Requirements. Cables shall be listed in accordance with 820.51(A) through (D).

(A) Type CATVP. Type CATVP community antenna television plenum cable shall be listed as being suitable for use in ducts, plenums, and other spaces used for environmental air and shall also be listed as having adequate fire-resistant and low smoke-producing characteristics.

> FPN: One method of defining low smoke-producing cables is by establishing an acceptable value of the smoke produced when tested in accordance with NFPA 262-1999, *Standard Method for Test for Flame Travel and Smoke of Wire and Cables for Use in Air-Handling Spaces,* to a maximum peak optical density of 0.5 and a maximum average optical density of 0.15. Similarly, one method of defining fire-resistant cables is by establishing maximum allowable flame travel distance of 1.52 m (5 ft) when tested in accordance with the same test.

(B) Type CATVR. Type CATVR community antenna television riser cable shall be listed as being suitable for use in a vertical run in a shaft or from floor to floor and shall also be listed as having fire-resistant characteristics capable of preventing the carrying of fire from floor to floor.

> FPN: One method of defining fire-resistant characteristics capable of preventing the carrying of fire from floor to floor is that the cables pass the requirements of ANSI/UL 1666-1997, *Standard Test for Flame*

Propagation Height of Electrical and Optical-Fiber Cable Installed Vertically in Shafts.

(C) Type CATV. Type CATV community antenna television cable shall be listed as being suitable for general-purpose CATV use, with the exception of risers and plenums, and shall also be listed as being resistant to the spread of fire.

> FPN: One method of defining *resistant to the spread of fire* is that the cables do not spread fire to the top of the tray in the vertical-tray flame test in ANSI/UL 1581-1991, *Reference Standard for Electrical Wires, Cables and Flexible Cords.*
>
> Another method of defining *resistant to the spread of fire* is for the damage (char length) not to exceed 1.5 m (4 ft 11 in.) when performing the vertical flame test for cables in cable trays, as described in CSA C22.2 No. 0.3-M-1985, *Test Methods for Electrical Wires and Cables.*

(D) Type CATVX. Type CATVX limited-use community antenna television cable shall be listed as being suitable for use in dwellings and for use in raceway and shall also be listed as being resistant to flame spread.

> FPN: One method of determining that cable is resistant to flame spread is by testing the cable to the VW-1 (vertical-wire) flame test in ANSI/UL 1581-1991, *Reference Standard for Electrical Wires, Cables and Flexible Cords.*

820.52 Installation of Cables and Equipment. Beyond the point of grounding, as defined in 820.33, the cable installation shall comply with 820.33(A) through (D).

(A) Separation from Other Conductors.

(1) In Raceways and Boxes.

(a) Other Circuits. Coaxial cables shall be permitted in the same raceway or enclosure with jacketed cables of any of the following:

(1) Class 2 and Class 3 remote-control, signaling, and power-limited circuits in compliance with Article 725
(2) Power-limited fire alarm systems in compliance with Article 760
(3) Nonconductive and conductive optical fiber cables in compliance with Article 770
(4) Communications circuits in compliance with Article 800
(5) Low-power network-powered broadband communications circuits in compliance with Article 830

(b) Electric Light, Power, Class 1, Non–Power-Limited Fire Alarm, and Medium Power Network-Powered Broadband Communications Circuits.

Coaxial cable shall not be placed in any raceway, compartment, outlet box, junction box, or other enclosures with conductors of electric light, power, Class 1, non–power-limited fire alarm, or medium power network-powered broadband communications circuits.

Exception No. 1: Where all of the conductors of electric light, power, Class 1, non–power-limited fire alarm, and medium power network-powered broadband communications circuits are separated from all of the coaxial cables by a barrier.

Exception No. 2: Power circuit conductors in outlet boxes, junction boxes, or similar fittings or compartments where such conductors are introduced solely for power supply to the coaxial cable system distribution equipment. The power circuit conductors shall be routed within the enclosure to maintain a minimum 6-mm (0.25-in.) separation from coaxial cables.

(2) Other Applications. Coaxial cable shall be separated at least 50 mm (2 in.) from conductors of any electric light, power, Class 1, non–power-limited fire alarm, or medium power network-powered broadband communications circuits.

Exception No. 1: Where either (1) all of the conductors of electric light, power, Class 1, non–power-limited fire alarm, and medium power network-powered broadband communications and circuits are in a raceway, or in metal-sheathed, metal-clad, nonmetallic-sheathed Type AC or Type UF cables, or (2) all of the coaxial cables are encased in raceway.

Exception No. 2: Where the coaxial cables are permanently separated from the conductors of electric light, power, Class 1, non–power-limited fire alarm, and medium power network-powered broadband communications circuits by a continuous and firmly fixed nonconductor, such as porcelain tubes or flexible tubing, in addition to the insulation on the wire.

(B) Equipment in Other Space Used for Environmental Air. Section 300.22(C) shall apply.

(C) Hybrid Power and Coaxial Cabling. The provisions of 780.6 shall apply for listed hybrid power and coaxial cabling in closed-loop and programmed power distribution.

(D) Support of Cables. Raceways shall be used for their intended purpose. Coaxial cables shall not be strapped, taped, or attached by any means to the exterior of any conduit or raceway as a means of support.

Exception: Overhead (aerial) spans of coaxial cables shall be permitted to be attached to the exterior of a raceway-type mast intended for the attachment and support of such cables.

820.53 Applications of Listed CATV Cables. CATV cables shall comply with the requirements of 820.53(A) through (D) or where cable substitutions are made as shown in Table 820.53.

(A) Plenum. Cables installed in ducts, plenums, and other spaces used for environmental air shall be Type CATVP. Abandoned cables shall not be permitted to remain. Types CATVP, CATVR, CATV, and CATVX cables installed in compliance with 300.22 shall be permitted.

(B) Riser. Cables installed in risers shall comply with any of the requirements of 820.53(B)(1) through (B)(3).

(1) Cables in Vertical Runs. Cables installed in vertical runs and penetrating more than one floor, or cables installed in vertical runs in a shaft, shall be Type CATVR. Floor penetrations requiring Type CATVR shall contain only cables suitable for riser or plenum use. Abandoned cables shall not be permitted to remain.

(2) Metal Raceways or Fireproof Shafts. Types CATV and CATVX cables shall be permitted to be encased in a metal raceway or located in a fireproof shaft having firestops at each floor.

(3) One- and Two-Family Dwellings. Types CATV and CATVX cables shall be permitted in one- and two-family dwellings.

FPN: See 820.53(A) for the firestop requirements for floor penetrations.

(C) Cable Trays. Cables installed in cable trays shall be Types CATVP, CATVR, and CATV.

Table 820.53 Coaxial Cable Uses and Permitted Substitutions

Cable Type	Use	References	Permitted Substitutions
CATVP	Coaxial plenum cable	820.53(A)	CMP
CATVR	Coaxial riser cable	820.53(B)	CATVP, CMP, CMR
CATV	Coaxial general-purpose cable	820.53(D)	CATVP, CMP, CATVR, CMR, CMG, CM
CATVX	Coaxial cable, limited use	820.53(D)	CATVP, CMP, CATVR, CMR, CATV, CMG, CM

Note: See Figure 820.53, Cable substitution hierarchy.
FPN: The substitute cables in Table 820.53 are only coaxial-type cables.

(D) Other Wiring Within Buildings. Cables installed in building locations other than the locations covered in 820.53(A) and (B) shall be with any of the requirements in 820.53(D)(1) through (5). Abandoned cables in hollow spaces shall not be permitted to remain.

(1) General. Type CATV shall be permitted.

(2) In Raceways. Type CATVX shall be permitted to be installed in a raceway.

(3) Nonconcealed Spaces. Type CATVX shall be permitted to be installed in nonconcealed spaces where the exposed length of cable does not exceed 3 m (10 ft).

(4) One- and Two-Family Dwellings. Type CATVX cables less than 10 mm (0.375 in.) in diameter shall be permitted to be installed in one- and two-family dwellings.

(5) Multifamily Dwellings. Type CATVX cables less than 10 mm (0.375 in.) in diameter shall be permitted to be installed in multifamily dwellings.

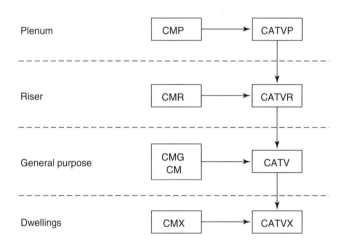

Type CATV—Community antenna television cables

Type CM—Communications cables

A ▶ B   Coaxial cable A shall be permitted to be used in place of coaxial cable B.

Figure 820.53 Cable substitution hierarchy.

Network-Powered Broadband Communications Systems

I. General

830.1 Scope. This article covers network-powered broadband communications systems that provide any combination of voice, audio, video, data, and interactive services through a network interface unit.

> FPN No. 1: A typical basic system configuration includes a cable supplying power and broadband signal to a network interface unit that converts the broadband signal to the component signals. Typical cables are coaxial cable with both broadband signal and power on the center conductor, composite metallic cable with a coaxial member for the broadband signal and a twisted pair for power, and composite optical fiber cable with a pair of conductors for power. Larger systems may also include network components such as amplifiers that require network power.

> FPN No. 2: See 90.2(B)(4) for installations of broadband communications systems that are not covered.

830.2 Definitions. See Article 100. For purposes of this article, the following additional definitions apply.

Abandoned Network-Powered Broadband Communications Cable. Installed network-powered broadband communications cable that is not terminated at equipment other than a connector and not identified for future use with a tag.

Block. A square or portion of a city, town, or village enclosed by streets, including the alleys so enclosed but not any street.

Exposed to Accidental Contact with Electrical Light or Power Conductors. A circuit in such a position that, in case of failure of supports or insulation, contact with another circuit may result.

Source: NFPA 70, *National Electrical Code®*, NFPA, Quincy, MA, 2002 edition.

Fault Protection Device. An electronic device that is intended for the protection of personnel and functions under fault conditions, such as network-powered broadband communications cable short or open circuit, to limit the current or voltage, or both, for a low-power network-powered broadband communications circuit and provide acceptable protection from electric shock.

Network Interface Unit (NIU). A device that converts a broadband signal into component voice, audio, video, data, and interactive services signals. The NIU provides isolation between the network power and the premises signal circuits. The NIU may also contain primary and secondary protectors.

Network-Powered Broadband Communications Circuit. The circuit extending from the communications utility's serving terminal or tap up to and including the NIU.

> FPN: A typical single-family network-powered communications circuit consists of a communications drop or communications service cable and an NIU and includes the communications utility's serving terminal or tap where it is not under the exclusive control of the communications utility.

Point of Entrance. The point within a building at which the cable emerges from an external wall, from a concrete floor slab, or from a rigid metal conduit or an intermediate metal conduit grounded to an electrode in accordance with 830.40(B).

Premises Wiring. The circuits located on the user side of the network interface unit.

830.3 Locations and Other Articles. Circuits and equipment shall comply with 830.3(A) through (D).

(A) Spread of Fire or Products of Combustion. Section 300.21 shall apply. The accessible portion of abandoned network-powered broadband communications cables shall not be permitted to remain.

(B) Ducts, Plenums, and Other Air-Handling Spaces. Section 300.22 shall apply, where installed in ducts or plenums or other spaces used for environmental air.

Exception: As permitted in 830.55(B).

(C) Installation and Use. Section 110.3(B) shall apply.

(D) Output Circuits. As appropriate for the services provided, the output circuits derived from the network interface unit shall comply with the requirements of the following:

(1) Installations of communications circuits — Article 800
(2) Installations of community antenna television and radio distribution circuits — Article 820

Exception: 830.30(B)(3) shall apply where protection is provided in the output of the NIU.

(3) Installations of optical fiber cables — Article 770
(4) Installations of Class 2 and Class 3 circuits — Article 725
(5) Installations of power-limited fire alarm circuits — Article 760

830.4 Power Limitations. Network-powered broadband communications systems shall be classified as having low or medium power sources as defined in Table 830.4.

830.5 Network-Powered Broadband Communications Equipment and Cables. Network-powered broadband communications equipment and cables shall be listed as suitable for the purpose.

Exception No. 1: This listing requirement shall not apply to community antenna television and radio distribution system coaxial cables that were installed prior to

Table 830.4   Limitations for Network-Powered Broadband Communications Systems

Network Power Source	Low	Medium
Circuit voltage, V_{max} (volts)[1]	0–100	0–150
Power limitation, VA_{max}(volt-amperes)[1]	250	250
Current limitation, I_{max} (amperes)[1]	$1000/V_{max}$	$1000/V_{max}$
Maximum power rating (volt-amperes)	100	100
Maximum voltage rating (volts)	100	150
Maximum overcurrent protection (amperes)[2]	$100/V_{max}$	NA

[1]V_{max}, I_{max}, and VA_{max} are determined with the current-limiting impedance in the circuit (not bypassed) as follows:

V_{max}—Maximum system voltage regardless of load with rated input applied.

I_{max}—Maximum system current under any noncapacitive load, including short circuit, and with overcurrent protection bypassed if used.

I_{max} limits apply after 1 minute of operation.

VA_{max}—Maximum volt-ampere output after 1 minute of operation regardless of load and overcurrent protection bypassed if used.

[2]Overcurrent protection is not required where the current-limiting device provides equivalent current limitation and the current-limiting device does not reset until power or the load is removed.

January 1, 2000, in accordance with Article 820 and are used for low-power network-powered broadband communications circuits. See 830.9.

Exception No. 2: Substitute cables for network-powered broadband communications cables shall be permitted as shown in Table 830.58.

(A) Listing and Marking. Listing and marking of network-powered broadband communications cables shall comply with 830.5(A)(1) or (A)(2).

(1) Type BMU, Type BM, and Type BMR Cables. Network-powered broadband communications medium power underground cable, Type BMU; network-powered broadband communications medium power cable, Type BM; and network-powered broadband communications medium power riser cable, Type BMR, shall be factory-assembled cables consisting of a jacketed coaxial cable, a jacketed combination of coaxial cable and multiple individual conductors, or a jacketed combination of an optical fiber cable and multiple individual conductors. The insulation for the individual conductors shall be rated for 300 volts minimum. Cables intended for outdoor use shall be listed as suitable for the application. Cables shall be marked in accordance with 310.11. Type BMU cables shall be jacketed and listed as being suitable for outdoor underground use. Type BM cables shall be listed as being suitable for general-purpose use, with the exception of risers and plenums, and shall also be listed as being resistant to the spread of fire. Type BMR cables shall be listed as being suitable for use in a vertical run in a shaft or from floor to floor and shall also be listed as having fire-resistant characteristics capable of preventing the carrying of fire from floor to floor.

> FPN No. 1: One method of defining *resistant to spread of fire* is that the cables do not spread fire to the top of the tray in the vertical tray flame test in ANSI/UL 1581-1991, *Reference Standard for Electrical Wires, Cables and Flexible Cords.* Another method of defining *resistant to the spread of fire* is for the damage (char length) not to exceed 1.5 m (4 ft 11 in.) when performing the CSA vertical flame test for cables in cable trays, as described in CSA C22.2 No. 0.3-M-1985, *Test Methods for Electrical Wires and Cables.*

> FPN No. 2: One method of defining fire-resistant characteristics capable of preventing the carrying of fire from floor to floor is that the cables pass the requirements of ANSI/UL 1666-1997, *Standard Test for Flame Propagation Height of Electrical and Optical-Fiber Cable Installed Vertically in Shafts.*

(2) Type BLU, Type BLX, and Type BLP Cables. Network-powered broadband communications low-power underground cable, Type BLU; limited use network-powered broadband communications low-power cable,

Type BLX; and network-powered broadband communications low-power plenum cable, Type BLP, shall be factory assembled cables consisting of a jacketed coaxial cable, a jacketed combination of coaxial cable and multiple individual conductors, or a jacketed combination of an optical fiber cable and multiple individual conductors. The insulation for the individual conductors shall be rated for 300 volts minimum. Cables intended for outdoor use shall be listed as suitable for the application. Cables shall be marked in accordance with 310.11. Type BLU cables shall be jacketed and listed as being suitable for outdoor underground use. Type BLX limited-use cables shall be listed as being suitable for use outside, for use in dwellings, and for use in raceways and shall also be listed as being resistant to flame spread. Type BLP cables shall be listed as being suitable for use in ducts, plenums, and other spaces for environmental air and shall also be listed as having adequate fire-resistant and low smoke-producing characteristics.

> FPN No. 1: One method of determining that cable is resistant to flame spread is by testing the cable to VW-1 (vertical-wire) flame test in ANSI/UL 1581-1991, *Reference Standard for Electrical Wires, Cables and Flexible Cords.*

> FPN No. 2: One method of defining low smoke-producing cable is by establishing an acceptable value of the smoke produced when tested in accordance with NFPA 262-1999, *Standard Method of Test for Flame Travel and Smoke of Wires and Cables for Use in Air-Handling Spaces,* to a maximum peak optical density of 0.5 and a maximum average optical density of 0.15. Similarly, one method of defining fire-resistant cables is by establishing maximum allowable flame travel distance of 1.52 m (5 ft) when tested in accordance with the same test.

830.6 Access to Electrical Equipment Behind Panels Designed to Allow Access. Access to electrical equipment shall not be denied by an accumulation of wires and cables that prevents removal of panels. including suspended ceiling panels.

830.7 Mechanical Execution of Work. Network-powered broadband communications circuits and equipment shall be installed in a neat and workmanlike manner. Cables installed exposed on the surface of ceiling and sidewalls shall be supported by the structural components of the building structure in such a manner that the cable is not damaged by normal building use. Such cables shall be attached to structural components by straps, staples, hangers, or similar fittings designed and installed so as not to damage the cable. The installation shall also conform with 300.4(D).

830.9 Hazardous (Classified) Locations. Network-powered broadband communications circuits and equipment installed in a location that is classified in accordance with Article 500 shall comply with the applicable requirements of Chapter 5.

II. Cables Outside and Entering Buildings

830.10 Entrance Cables. Cables installed outdoors shall be listed as suitable for the application. In addition, network-powered broadband communications cables located outside and entering buildings shall comply with 830.10(A) and (B).

(A) Medium Power Circuits. Medium power network-powered broadband communications circuits located outside and entering buildings shall be installed using Type BMU, Type BM, or Type BMR network-powered broadband communications medium power cables.

(B) Low-Power Circuits. Low-power network-powered broadband communications circuits located outside and entering buildings shall be installed using Type BLU or Type BLX low-power network-powered broadband communications cables. Cables shown in Table 830.58 shall be permitted to substitute.

Exception: Outdoor community antenna television and radio distribution system coaxial cables installed prior to January 1, 2000, and installed in accordance with Article 820, shall be permitted for low-power-type, network-powered broadband communications circuits.

830.11 Aerial Cables. Aerial powered broadband communications cables shall comply with 830.11(A) through (I).

   FPN: For additional information regarding overhead wires and cables, see ANSI C2-1997, *National Electric Safety Code*, Part 2, Safety Rules For Overhead Lines.

(A) On Poles. Where practicable, network-powered broadband communications cables on poles shall be located below the electric light, power, Class 1, or non–power-limited fire alarm circuit conductors and shall not be attached to a cross-arm that carries electric light or power conductors.

(B) Climbing Space. The climbing space through network-powered broadband communications cables shall comply with the requirements of 225.14(D).

(C) Lead-in Clearance. Lead-in or aerial-drop network-powered broadband communications cables from a pole or other support, including the point of initial attachment to a building or structure, shall be kept away from electric light, power, Class 1, or non–power-limited fire alarm circuit conductors so as to avoid the possibility of accidental contact.

Exception: Where proximity to electric light, power, Class 1, or non–power-limited fire alarm circuit service conductors cannot be avoided, the installation shall be such as to provide clearances of not less than 300 mm (12 in.) from light, power, Class 1, or non–power-limited fire alarm circuit service drops. The clearance requirement shall apply to all points along the drop, and it shall increase to 1.02 m (40 in.) at the pole.

(D) Clearance from Ground. Overhead spans of network-powered broadband communication cables shall conform to not less than the following:

(1) 2.9 m (9.5 ft) — above finished grade, sidewalks, or from any platform or projection from which they might be reached and accessible to pedestrians only
(2) 3.5 m (11.5 ft) — over residential property and driveways, and those commercial areas not subject to truck traffic
(3) 4.7 m (15.5 ft) — over public streets, alleys, roads, parking areas subject to truck traffic, driveways on other than residential property, and other land traversed by vehicles such as cultivated, grazing, forest, and orchard

> FPN: These clearances have been specifically chosen to correlate with ANSI C2-1997, *National Electrical Safety Code*, Table 232-1, which provides for clearances of wires, conductors, and cables above ground and roadways, rather than using the clearances referenced in 225.18. Because Article 800 and Article 820 have had no required clearances, the communications industry has used the clearances from the NESC for their installed cable plant.

(E) Over Pools. Clearance of network-powered broadband communications cable in any direction from the water level, edge of pool, base of diving platform, or anchored raft shall comply with those clearances in 680.8.

(F) Above Roofs. Network-powered broadband communications cables shall have a vertical clearance of not less than 2.5 m (8 ft) from all points of roofs above which they pass.

Exception No. 1: Auxiliary buildings such as garages and the like.

Exception No. 2: A reduction in clearance above only the overhanging portion of the roof to not less than 450 mm (18 in.) shall be permitted if (1) not more than 1.2 m (4 ft) of the broadband communications drop cables pass above the roof overhang, and (2) they are terminated at a through-the-roof raceway or support.

Exception No. 3: Where the roof has a slope of not less than 100 mm (4 in.) in 300 mm (12 in.), a reduction in clearance to not less than 900 mm (3 ft) shall be permitted.

(G) Final Spans. Final spans of network-powered broadband communications cables without an outer jacket shall be permitted to be attached to the building, but they shall be kept not less than 900 mm (3 ft) from windows that are designed to be opened, doors, porches, balconies, ladders, stairs, fire escapes, or similar locations.

Exception: Conductors run above the top level of a window shall be permitted to be less than the 900-mm (3-ft) requirement above.

Overhead network-powered broadband communications cables shall not be installed beneath openings through which materials may be moved, such as openings in farm and commercial buildings, and shall not be installed where they will obstruct entrance to these building openings.

(H) Between Buildings. Network-powered broadband communications cables extending between buildings and also the supports or attachment fixtures shall be acceptable for the purpose and shall have sufficient strength to withstand the loads to which they may be subjected.

Exception: Where a network-powered broadband communications cable does not have sufficient strength to be self-supporting, it shall be attached to a supporting messenger cable that, together with the attachment fixtures or supports, shall be acceptable for the purpose and shall have sufficient strength to withstand the loads to which they may be subjected.

(I) On Buildings. Where attached to buildings, network-powered broadband communications cables shall be securely fastened in such a manner that they are separated from other conductors in accordance with 830.11(I)(1) through (I)(4).

(1) Electric Light or Power. The network-powered broadband communications cable shall have a separation of at least 100 mm (4 in.) from electric light, power, Class 1, or non–power-limited fire alarm circuit conductors not in raceway or cable, or be permanently separated from conductors of the other system by a continuous and firmly fixed nonconductor in addition to the insulation on the wires.

(2) Other Communications Systems. Network-powered broadband communications cables shall be installed so that there will be no unnecessary interference in the maintenance of the separate systems. In no case shall the conductors, cables, messenger strand, or equipment of one system cause abrasion to the conductors, cables, messenger strand, or equipment of any other system.

(3) Lightning Conductors. Where practicable, a separation of at least 1.8 m (6 ft) shall be maintained between any network-powered broadband communications cable and lightning conductors.

(4) Protection from Damage. Network-powered broadband communications cables attached to buildings and located within 2.5 m (8 ft) of finished grade shall be protected by enclosures, raceways, or other approved means.

Exception: A low-power network-powered broadband communications circuit that is equipped with a listed fault protection device, appropriate to the network-powered broadband communications cable used, and located on the network side of the network-powered broadband communications cable being protected.

830.12 Underground Circuits Entering Buildings.

(A) Underground Systems. Underground network-powered broadband communications cables in a duct, pedestal, handhole, or manhole that contains electric light, power conductors, non–power-limited fire alarm circuit conductors, or Class 1 circuits shall be in a section permanently separated from such conductors by means of a suitable barrier.

(B) Direct-Buried Cables and Raceways. Direct-buried network-powered broadband communications cables shall be separated at least 300 mm (12 in.) from conductors of any light, power, non–power-limited fire alarm circuit conductors or Class 1 circuit.

Exception No. 1: Where electric service conductors or network-powered broadband communications cables are installed in raceways or have metal cable armor.

Exception No. 2: Where electric light or power branch-circuit or feeder conductors, non–power-limited fire alarm circuit conductors, or Class 1 circuit conductors are installed in a raceway or in metal-sheathed, metal-clad, or Type UF or Type USE cables; or the network-powered broadband communications cables have metal cable armor or are installed in a raceway.

(C) Mechanical Protection. Direct-buried cable, conduit, or other raceways shall be installed to meet the minimum cover requirements of Table 830.12. In addition, direct-buried cables emerging from the ground shall be protected by enclosures, raceways, or other approved means extending from the minimum cover distance required by Table 830.12 below grade to a point at least 2.5 m (8 ft) above finished grade. In no case shall the protection be required to exceed 450 mm (18 in.) below finished grade. Type BMU and BLU direct-buried cables emerging from the ground shall be installed in rigid metal conduit, intermediate metal conduit, rigid nonmetallic conduit, or other approved means extending from the minimum cover distance required by Table 830.12 below grade to the point of entrance.

Exception: A low-power network-powered broadband communications circuit that is equipped with a listed fault protection device, appropriate to the network-powered broadband communications cable used, and located on the network side of the network-powered broadband communications cable being protected.

Table 830.12 Network-Powered Broadband Communications Systems Minimum Cover Requirements (*Cover* is the shortest distance measured between a point on the top surface of any direct-buried cable, conduit, or other raceway and the top surface of finished grade, concrete, or similar cover.)

Location of Wiring Method or Circuit	Direct Burial Cables		Rigid Metal Conduit or Intermediate Metal Conduit		Nonmetallic Raceways Listed for Direct Burial; Without Concrete Encasement or Other Approved Raceways	
	mm	in.	mm	in.	mm	in.
All locations not specified below	450	18	150	6	300	12
In trench below 50-mm (2-in.) thick concrete or equivalent	300	12	150	6	150	6
Under a building (in raceway only)	0	0	0	0	0	0
Under minimum of 100-mm (4-in.) thick concrete exterior slab with no vehicular traffic and the slab extending not less than 150 mm (6 in.) beyond the underground installation	300	12	100	4	100	4
One- and two-family dwelling driveways and outdoor parking areas and used only for dwelling-related purposes	300	12	300	12	300	12

Notes:

1. Raceways approved for burial only where concrete encased shall require a concrete envelope not less than 50 mm (2 in.) thick.

2. Lesser depths shall be permitted where cables rise for terminations or splices or where access is otherwise required.

3. Where solid rock is encountered, all wiring shall be installed in metal or nonmetallic raceway permitted for direct burial. The raceways shall be covered by a minimum of 50 mm (2 in.) of concrete extending down to rock.

4. Low-power network-powered broadband communications circuits using directly buried community antenna television and radio distribution systecoaxial cables that were installed outside and entering buildings prior to January 1, 2000, in accordance with Article 820 shall be permitted where buried to a minimum depth of 300 mm (12 in.).

(D) Pools. Cables located under the pool or within the area extending 1.5 m (5 ft) horizontally from the inside wall of the pool shall meet those clearances and requirements specified in 680.10.

III. Protection

830.30 Primary Electrical Protection.

(A) Application. Primary electrical protection shall be provided on all network-powered broadband communications conductors that are neither grounded nor interrupted and are run partly or entirely in aerial cable not confined within a block. Also, primary electrical protection shall be provided on all aerial or underground network-powered broadband communications conductors that are neither grounded nor interrupted and are located within the block containing the building served so as to be exposed to lightning or accidental contact with electric light or power conductors operating at over 300 volts to ground.

Exception: Where electrical protection is provided on the derived circuit(s) (output side of the NIU) in accordance with 830.30(B)(3).

> FPN No. 1: On network-powered broadband communications conductors not exposed to lightning or accidental contact with power conductors, providing primary electrical protection in accordance with this article helps protect against other hazards, such as ground potential rise caused by power fault currents, and above-normal voltages induced by fault currents on power circuits in proximity to the network-powered broadband communications conductors.

> FPN No. 2: Network-powered broadband communications circuits are considered to have a lightning exposure unless one or more of the following conditions exist:

> (1) Circuits in large metropolitan areas where buildings are close together and sufficiently high to intercept lightning.
> (2) Areas having an average of five or fewer thunderstorm days per year and earth resistivity of less than 100 ohm-meters. Such areas are found along the Pacific coast.

(1) Fuseless Primary Protectors. Fuseless-type primary protectors shall be permitted where power fault currents on all protected conductors in the cable are safely limited to a value no greater than the current-carrying capacity of the primary protector and of the primary protector grounding conductor.

(2) Fused Primary Protectors. Where the requirements listed in 830.30(A)(1) are not met, fused-type primary protectors shall be used. Fused-type primary protectors shall consist of an arrester connected between each conductor to be protected and ground, a fuse in series with each conductor to be protected, and an appropriate mounting arrangement. Fused primary protector terminals shall be marked to indicate line, instrument, and ground, as applicable.

(B) Location. The location of the primary protector, where required, shall comply with (1), (2), or (3):

(1) A listed primary protector shall be applied on each network-powered broadband communications cable external to and on the network side of the network interface unit.
(2) The primary protection function shall be an integral part of and contained in the network interface unit. The network interface unit shall be listed for the purpose and shall have an external marking indicating that it contains primary electrical protection.
(3) The primary protector(s) shall be provided on the derived circuit(s) (output side of the NIU), and the combination of the NIU and the protector(s) shall be listed for the purpose.

A primary protector, whether provided integrally or external to the network interface unit, shall be located as close as practicable to the point of entrance.

For purposes of this section, a network interface unit and any externally provided primary protectors located at mobile home service equipment located in sight from and not more than 9.0 m (30 ft) from the exterior wall of the mobile home it serves, or at a mobile home disconnecting means grounded in accordance with 250.32 and located in sight from and not more than 9.0 m (30 ft) from the exterior wall of the mobile home it serves, shall be considered to meet the requirements of this section.

FPN: Selecting a network interface unit and primary protector location to achieve the shortest practicable primary protector grounding conductor helps limit potential differences between communications circuits and other metallic systems.

(C) Hazardous (Classified) Locations. The primary protector or equipment providing the primary protection function shall not be located in any hazardous (classified) location as defined in Article 500 or in the vicinity of easily ignitible material.

Exception: As permitted in 501.14, 502.14, and 503.12.

830.33 Grounding or Interruption of Metallic Members of Network-Powered Broadband Communications Cables. The shields of network-powered broadband communications cables used for communications or powering shall be grounded at the building as close as practicable to the point of entrance or attachment of the NIU. Metallic cable members not used for communications or powering shall be grounded or interrupted by an insulating joint or equivalent device as close as practicable to the point of entrance or attachment of the NIU.

For purposes of this section, grounding or interruption of network-powered broadband communications cable metallic members installed at mobile home service equipment located in sight from and no more than 9.0 m (30 ft) from the exterior wall of the mobile home it serves, or at a mobile home disconnecting means grounded in accordance with 250.32 and located in sight from and not more than 9.0 m (30 ft) from the exterior wall of the mobile home it serves, shall be considered to meet the requirements of this section.

> FPN: Selecting a grounding location to achieve the shortest practicable grounding conductor helps limit potential differences between the network-powered broadband communications circuits and other metallic systems.

IV. Grounding Methods

830.40 Cable, Network Interface Unit, and Primary Protector Grounding. Network interface units containing protectors, NIUs with metallic enclosures, primary protectors, and the metallic members of the network-powered broadband communications cable that are intended to be grounded shall be grounded as specified in 830.40(A) through (D).

(A) Grounding Conductor.

(1) Insulation. The grounding conductor shall be insulated and shall be listed as suitable for the purpose.

(2) Material. The grounding conductor shall be copper or other corrosion-resistant conductive material, stranded or solid.

(3) Size. The grounding conductor shall not be smaller than 14 AWG and shall have a current-carrying capacity approximately equal to that of the grounded metallic member(s) and protected conductor(s) of the network-powered broadband communications cable. The grounding conductor shall not be required to exceed 6 AWG.

(4) Length. The grounding conductor shall be as short as practicable. In one-family and multifamily dwellings, the grounding conductor shall be as short as permissible, not to exceed 6.0 m (20 ft) in length.

Exception: In one- and two-family dwellings where it is not practicable to achieve an overall maximum grounding conductor length of 6.0 m (20 ft), a separate communications ground rod meeting the minimum dimensional criteria of 830.40(B)(2)(2) shall be driven, and the grounding conductor connected to the communications ground rod in accordance with 830.40(C). The communications ground rod shall be bonded to the power grounding electrode system in accordance with 830.40(D).

(5) Run in Straight Line. The grounding conductor shall be run to the grounding electrode in as straight a line as practicable.

(6) Physical Protection. Where subject to physical damage, the grounding conductor shall be adequately protected. Where the grounding conductor is run in a metal raceway, both ends of the raceway shall be bonded to the grounding conductor or the same terminal or electrode to which the grounding conductor is connected.

(B) Electrode. The grounding conductor shall be connected as follows.

(1) In Buildings or Structures with Grounding Means. To the nearest accessible location on the following:

(1) The building or structure grounding electrode system as covered in 250.50;
(2) The grounded interior metal water piping system, within 1.5 m (5 ft) from its point of entrance to the building, as covered in 250.52;
(3) The power service accessible means external to enclosures as covered in 250.94;
(4) The metallic power service raceway;
(5) The service equipment enclosure;
(6) The grounding electrode conductor or the grounding electrode metal enclosure; or
(7) The grounding conductor or the grounding electrode of a building or structure disconnecting means that is grounded to an electrode as covered in 250.32.

For purposes of this section, the mobile home service equipment or the mobile home disconnecting means, as described in 830.33, shall be considered accessible.

(2) In Buildings or Structures Without Grounding Means. If the building or structure served has no grounding means, as described in (B)(1):

(1) To any one of the individual electrodes described in 250.52(A)(1), (2), (3), (4); or
(2) If the building or structure served has no grounding means, as described in 830.40(B)(1) or (B)(2)(1), to an effectively grounded metal structure

or to a ground rod or pipe not less than 1.5 m (5 ft) in length and 12.7 mm (½ in.) in diameter, driven, where practicable, into permanently damp earth and separated from lightning conductors as covered in 800.13 and at least 1.8 m (6 ft) from electrodes of other systems. Steam or hot water pipes or lightning-rod conductors shall not be employed as electrodes for protectors, NIUs with integral protection, grounded metallic members, NIUs with metallic enclosures, and other equipment.

(C) Electrode Connection. Connections to grounding electrodes shall comply with 250.70. Connectors, clamps, fittings, or lugs used to attach grounding conductors and bonding jumpers to grounding electrodes or to each other that are to be concrete encased or buried in the earth shall be suitable for its application.

(D) Bonding of Electrodes. A bonding jumper not smaller than 6 AWG copper or equivalent shall be connected between the network-powered broadband communications system grounding electrode and the power grounding electrode system at the building or structure served where separate electrodes are used.

Exception: At mobile homes as covered in 830.42.

FPN No. 1: See 250.60 for use of lightning rods.

FPN No. 2: Bonding together of all separate electrodes limits potential differences between them and between their associated wiring systems.

830.42 Bonding and Grounding at Mobile Homes.

(A) Grounding. Where there is no mobile home service equipment located in sight from and not more than 9.0 m (30 ft) from the exterior wall of the mobile home it serves, or there is no mobile home disconnecting means grounded in accordance with 250.32 and located within sight from and not more than 9.0 m (30 ft) from the exterior wall of the mobile home it serves, the network-powered broadband communications cable, network interface unit, and primary protector ground shall be installed in accordance with 830.40(B)(2).

(B) Bonding. The network-powered broadband communications cable grounding terminal, network interface unit grounding terminal, if present, and primary protector grounding terminal shall be bonded together with a copper bonding conductor not smaller than 12 AWG. The network-powered broadband communications cable grounding terminal, network interface unit grounding terminal, primary protector grounding terminal, or the grounding electrode shall be bonded to the metal frame or available grounding terminal of the mobile home with a copper bonding conductor not smaller than 12 AWG under any of the following conditions:

(1) Where there is no mobile home service equipment or disconnecting means as in (A)
(2) Where the mobile home is supplied by cord and plug

V. Wiring Methods Within Buildings

830.54 Medium Power Network-Powered Broadband Communications System Wiring Methods. Medium power network-powered broadband communications systems shall be installed within buildings using listed Type BM or Type BMR, network-powered broadband communications medium power cables.

(A) Ducts, Plenums, and Other Air-Handling Spaces. Section 300.22 shall apply.

(B) Riser. Cables installed in vertical runs and penetrating more than one floor, or cables installed in vertical runs in a shaft, shall be Type BMR. Floor penetrations requiring Type BMR shall contain only cables suitable for riser or plenum use.

Exception No. 1: Type BM cables encased in metal raceway or located in a fireproof shaft that has firestops at each floor.

Exception No. 2: Type BM cables in one- and two-family dwellings.

(C) Other Wiring. Cables installed in locations other than the locations covered in 830.54(A) and (B) shall be Type BM.

Exception: Type BMU cable where the cable enters the building from the outside and is run in rigid metal conduit or intermediate metal conduit, and such conduits are grounded to an electrode in accordance with 830.40(B).

830.55 Low-Power Network-Powered Broadband Communications System Wiring Methods. Low-power network-powered broadband communications systems shall comply with any of the requirements of 830.55(A) through (D).

(A) In Buildings. Low-power network-powered broadband communications systems shall be installed within buildings using listed Type BLX or Type BLP network-powered broadband communications low power cables.

(B) Ducts, Plenums, and Other Air-Handling Spaces. Cables installed in ducts, plenums, and other spaces used for environmental air shall be Type BLP. Abandoned cables shall not be permitted to remain. Type BLX cable installed in compliance with 300.22 shall be permitted.

(C) Riser. Cables installed in risers shall comply with any of the requirements in 830.55(C)(1), (C)(2), or (C)(3).

(1) Cables in Vertical Runs. Cables installed in vertical runs and penetrating more than one floor, or cables installed in vertical runs in a shaft, shall be Type BLP or BMR. Floor penetrations requiring Type BMR shall contain only cables suitable for riser or plenum use. Abandoned cables shall not be permitted to remain.

(2) Metal Raceways or Fireproof Shafts. Type BLX cables shall be permitted to be encased in a metal raceway or located in a fireproof shaft having firestops at each floor.

(3) One- and Two-Family Dwellings. Type BLX cables less than 10 mm (0.375 in.) in diameter shall be permitted in one- and two-family dwellings.

(D) Other Wiring. Cables installed in locations other than the locations covered in 830.55(A), (B), and (C) shall comply with the requirements of 830.55(D)(1) through (D)(5).

(1) General. Type BLP or BM shall be permitted.

(2) In Raceways. Type BLX shall be permitted to be installed in a raceway.

(3) Type BLU Cable. Type BLU cable entering the building from outside shall be permitted to be run in rigid metal conduit or intermediate metal conduit. Such conduits shall be grounded to an electrode in accordance with 830.40(B).

(4) One- and Two-Family Dwellings. Type BLX cable less than 10 mm (0.375 in.) in diameter shall be permitted to be installed in one- and two-family dwellings.

(5) Type BLX Cable. Type BLX cable entering the building from outside and terminated at a grounding block or a primary protection location shall be permitted to be installed, provided that the length of cable within the building does not exceed 15 m (50 ft).

> FPN: This provision limits the length of Type BLX cable to 15 m (50 ft), while 830.30(B) requires that the primary protector, or NIU with integral protection, be located as close as practicable to the point at which the cable enters the building. Therefore, in installations requiring a primary protector, or NIU with integral protection, Type BLX cable may not be permitted to extend 15 m (50 ft) into the building if it is practicable to place the primary protector closer than 15 m (50 ft) to the entrance point.

830.56 Protection Against Physical Damage. Section 300.4 shall apply.

830.57 Bends. Bends in network broadband cable shall be made so as not to damage the cable.

830.58 Installation of Network-Powered Broadband Communications Cables and Equipment. Cable and equipment installations within buildings shall comply with 830.58(A) through (E), as applicable.

(A) Separation of Conductors.

(1) In Raceways and Enclosures.

(a) Low and Medium Power Network-Powered Broadband Communications Circuit Cables. Low and medium power network-powered broadband communications cables shall be permitted in the same raceway or enclosure.

(b) Low Power Network-Powered Broadband Communications Circuit Cables. Low power network-powered broadband communications cables shall be permitted in the same raceway or enclosure with jacketed cables of any of the following circuits:

(1) Class 2 and Class 3 remote-control, signaling, and power-limited circuits in compliance with Article 725
(2) Power-limited fire alarm systems in compliance with Article 760
(3) Communications circuits in compliance with Article 800
(4) Nonconductive and conductive optical fiber cables in compliance with Article 770
(5) Community antenna television and radio distribution systems in compliance with Article 820

(c) Medium Power Network-Powered Broadband Communications Circuit Cables. Medium power network-powered broadband communications cables shall not be permitted in the same raceway or enclosure with conductors of any of the following circuits:

(1) Class 2 and Class 3 remote-control, signaling, and power-limited circuits in compliance with Article 725
(2) Power-limited fire alarm systems in compliance with Article 760
(3) Communications circuits in compliance with Article 800
(4) Conductive optical fiber cables in compliance with Article 770
(5) Community antenna television and radio distribution systems in compliance with Article 820

(d) Electric Light, Power, Class 1, Non–Powered Broadband Communications Circuit Cables. Network-powered broadband communications cable shall not be placed in any raceway, compartment, outlet box, junction box, or similar fittings with conductors of electric light, power, Class 1, or non–power-limited fire alarm circuit cables.

Exception No. 1: Where all of the conductors of electric light, power, Class 1, non–power-limited fire alarm circuits are separated from all of the network-powered broadband communications cables by a barrier.

Exception No. 2: Power circuit conductors in outlet boxes, junction boxes, or similar fittings or compartments where such conductors are introduced solely for power supply to the network-powered broadband communications system distribution equipment. The power circuit conductors shall be routed within the enclosure to maintain a minimum 6-mm (0.25-in.) separation from network-powered broadband communications cables.

(2) Other Applications. Network-powered broadband communications cable shall be separated at least 50 mm (2 in.) from conductors of any electric light, power, Class 1, and non–power-limited fire alarm circuits.

Exception No. 1: Where either (1) all of the conductors of electric light, power, Class 1, and non–power-limited fire alarm circuits are in a raceway, or in metal-sheathed, metal-clad, nonmetallic-sheathed, Type AC, or Type UF cables, or (2) all of the network-powered broadband communications cables are encased in raceway.

Exception No. 2: Where the network-powered broadband communications cables are permanently separated from the conductors of electric light, power, Class 1, and non–power-limited fire alarm circuits by a continuous and firmly fixed nonconductor, such as porcelain tubes or flexible tubing, in addition to the insulation on the wire.

(B) Spread of Fire or Products of Combustion. Installations in hollow spaces, vertical shafts, and ventilation or air-handling ducts shall be made so that the possible spread of fire or products of combustion will not be substantially increased. Openings around penetrations through fire-resistance–rated walls, partitions, floors, or ceilings shall be firestopped using approved methods to maintain the fire resistance rating.

(C) Equipment in Other Space Used for Environmental Air. Section 300.22(C) shall apply.

(D) Support of Conductors. Raceways shall be used for their intended purpose. Network-powered broadband communications cables shall not be strapped, taped, or attached by any means to the exterior of any conduit or raceway as a means of support.

(E) Cable Substitutions. The substitutions for network-powered broadband cables listed in Table 830.58 shall be permitted. All cables in Table 830.58, other than network-powered broadband cables, shall be coaxial cables.

Table 830.58 Cable Substitutions

Cable Type	Permitted Cable Substitutions
BM	BMR
BLP	CMP, CL3P
BLX	CMP, CL3P, CMR, CL3R, CMG, CM, CL3, CMX, CL3X, BMR, BM, BLP

Bibliography

Bunker, M. W., and W. D. Moore, eds., *National Fire Alarm Code® Handbook*, 3rd ed., National Fire Protection Association, Quincy, MA, 1999.

Earley, M. W., M. C. Ode, J. S. Sargent, D. Strube, and N. Williams, *NEC® Changes 1999*, National Fire Protection Association, Quincy, MA, 1999.

Earley, M. W., J. V. Sheehan, and J. M. Caloggero, *National Electrical Code® Handbook*, 8th ed., National Fire Protection Association, Quincy, MA, 1999.

Earley, M. W., J. V. Sheehan, J. S. Sargent, J. M. Caloggero, and T. Croushore, *National Electrical Code® Handbook*, 9th ed., National Fire Protection Association, Quincy, MA, 2002.

Jones, R. A., and J. G. Jones, *Electrical Safety in the Workplace*, National Fire Protection Association, Quincy, MA, 2000.

NFPA 70, *National Electrical Code®*, National Fire Protection Association, Quincy, MA, 2002.

NFPA 72®, National Fire Alarm Code®, National Fire Protection Association, Quincy, MA, 1999.

NFPA 101®, *Life Safety Code®*, National Fire Protection Association, Quincy, MA, 2000.

Schram, P. J., and M. W. Earley, *Electrical Installations in Hazardous Locations*, National Fire Protection Association, Quincy, MA, 1997.

UL 13, *Standard for Safety for Power-Limited Circuit Cables*, Underwriters Laboratories, Northbrook, IL, 1996.

UL 444, *Standard for Safety for Communications Cables*, Underwriters Laboratories, Northbrook, IL, 1994.

UL 497, *Standard for Safety for Protectors for Paired-Conductor Communications Circuits*, Underwriters Laboratories, Northbrook, IL, 2001.

UL 1012, *Standard for Safety for Power Units Other Than Class 2*, Underwriters Laboratories, Northbrook, IL, 1994.

UL 1310, *Standard for Safety for Class 2 Power Units,* Underwriters Laboratories, Northbrook, IL, 1994.

UL 1424, *Standard for Safety for Cables for Power-Limited Fire-Alarm Circuits,* Underwriters Laboratories, Northbrook, IL, 1996.

UL 1950, *Standard for Safety of Information Technology Equipment, Including Electric Business Machines,* Underwriters Laboratories, Northbrook, IL, 1998.

Index

About the Author

Noel Williams is a licensed master electrician in the states of Utah, Colorado, and Wyoming. He is ICBO and IAEI certified as an electrical inspector. He has supervised and managed electrical construction projects for more than 20 years. For over 10 years he was chairman of the Electricians Licensing Board for the state of Utah and is currently chairman of the *NEC®* Advisory Committee for the Utah Uniform Building Code Commission. Mr. Williams serves on the executive committee of the Utah chapter of IAEI. Now a consultant and contributing editor for *CEE News,* he also taught electrical apprentices at Salt Lake Community College for 6 years and has developed and taught electrical courses for electricians and inspectors for more than 15 years. In addition to coauthoring two recent National Fire Protection Association (NFPA) publications, *1999 NEC Changes* and *Electrical Inspection Manual with Checklists,* Mr. Williams was lead technical developer of NFPA's *NEC* seminars and has served as a seminar instructor for over 10 years.

Skeletons

Meish Goldish

Contents

An Inside Look

Imagine that you have X-ray vision. If you look at any animal, you can see inside it. What you see will depend on the kind of animal you are looking at.

If you look inside a mouse, a snake, a person, or many other animals, you will see a **skeleton**. Some other animals wear their skeletons on the outside, and some have no skeleton at all. Get set for an amazing inside look!

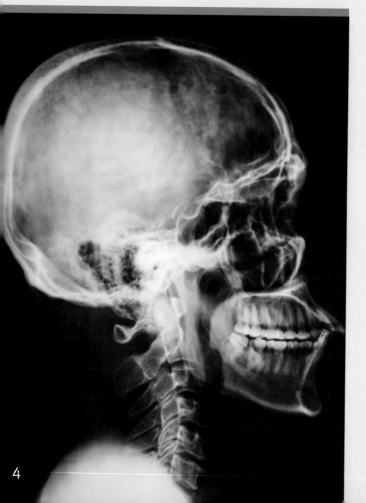

The skull is an important part of a skeleton. This is an X-ray of a skull.

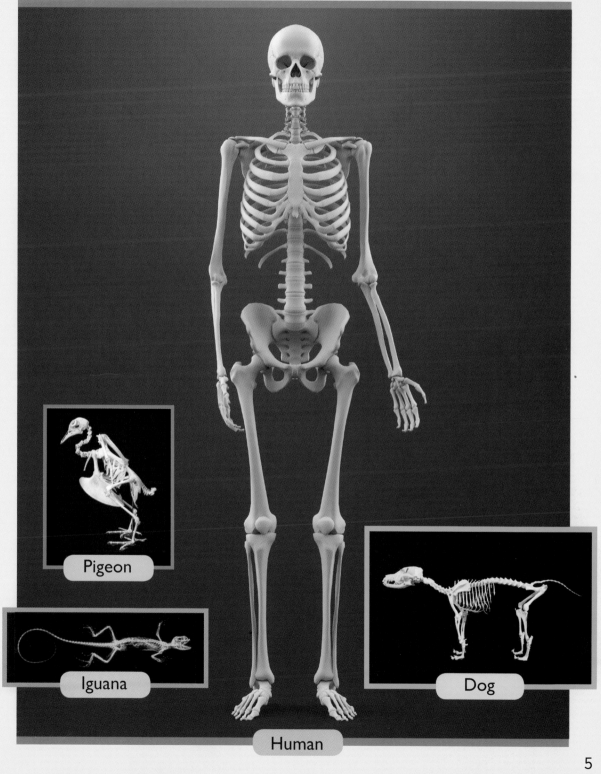

Pigeon

Iguana

Dog

Human

What Is a Skeleton?

We usually think of a skeleton as an animal's framework of bones. Many kinds of animals, including humans, have skeletons made of bones. Other animals have different kinds of skeletons, which you can read about later in this book.

Bones have different shapes. Some are like broad, flat plates. Others are hollow tubes with thick walls. Bones are also different sizes. The smallest bone in your body is a bone in your ear that is smaller than a grain of rice. Your biggest bone is your thighbone.

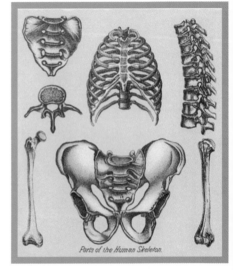

Parts of the Human Skeleton.

Medical books long ago had drawings of bones instead of photos.

A kangaroo has long, strong bones in its hind legs.

Bones are made of living tissue that grows and changes. Like other parts of the body, bones have blood vessels and nerves. But unlike other body tissues, bones are filled with a **mineral** that makes them hard and strong. That's why bones last longer than other parts of an animal's body after it dies.

Inside each bone is a thick, pasty material called **marrow**. Blood cells are formed in the marrow.

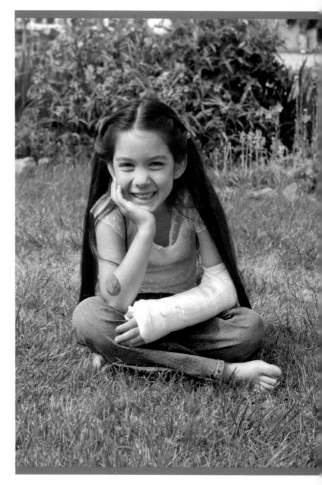

If you break a bone, it will heal. The broken pieces will grow back together.

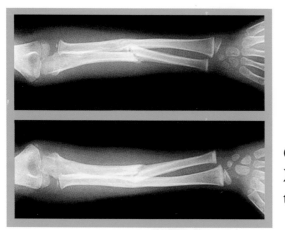

Can you tell which of these X-rays shows a broken bone that is healing?

7

What Does a Skeleton Do?

An animal's skeleton supports its body and gives the animal its particular shape. The skeleton also protects an animal's soft organs, such as its heart and lungs. Most important, skeletons help animals move.

When you move, your bones and muscles work together.

Cheetahs move faster than any other land animals. A cheetah can run up to 70 miles per hour!

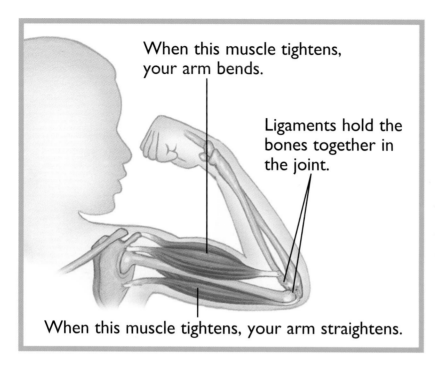

When this muscle tightens, your arm bends.

Ligaments hold the bones together in the joint.

Every joint in your body has at least one pair of muscles to move it.

When this muscle tightens, your arm straightens.

Bones are connected to each other at the ends by bands of tough tissue called **ligaments**. The connection between two bones is called a **joint**. Some joints, such as your knee and elbow, are like hinges that let the bones move back and forth. Your shoulder and hip joints let the bones turn in a circle.

Every muscle in an animal's body is attached to two different bones. When muscles tighten, they move the bones they are attached to.

Animals with Backbones

One of the most important parts of your skeleton is your backbone, which is also called your spine. Your backbone supports your body and protects your delicate spinal cord. Your spinal cord carries messages from your brain to all the other parts of your body. Your brain might send a message to your arm to throw a ball, or to your legs when you want to run. Without your backbone and spinal cord, you would not be able to stand, sit, or move.

Your backbone is not just one bone. It is made up of 33 small bones, called **vertebrae**. Animals with backbones are called **vertebrates**. All mammals, including humans, are vertebrates. So are birds, amphibians, reptiles, and fish.

All mammals have seven neck bones in their spine. Why do you think a giraffe's neck is so much longer than yours?

Human Skeleton

The words in italics are the names scientists use for the bones in the human skeleton.

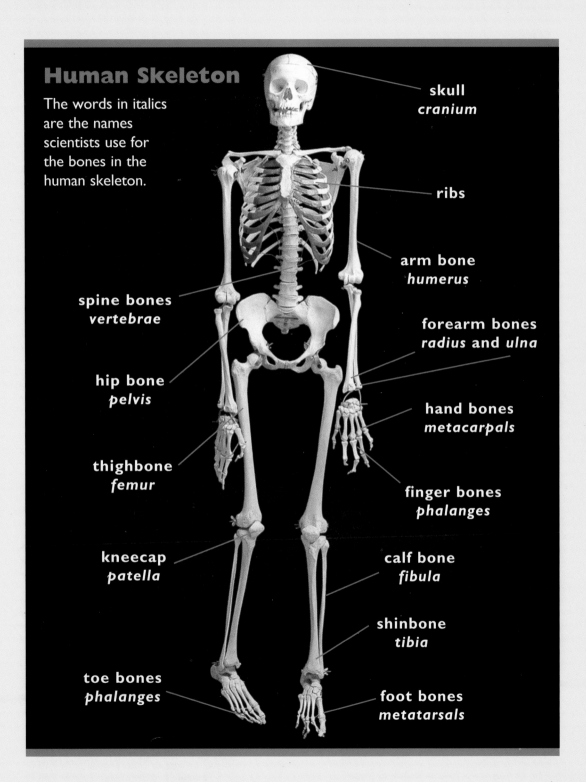

skull
cranium

ribs

arm bone
humerus

spine bones
vertebrae

forearm bones
radius and *ulna*

hip bone
pelvis

hand bones
metacarpals

thighbone
femur

finger bones
phalanges

kneecap
patella

calf bone
fibula

shinbone
tibia

toe bones
phalanges

foot bones
metatarsals

Slither, Leap, Swim

Animals that move in different ways have different kinds of skeletons. A snake is a vertebrate that moves by curving its body from side to side. This movement pushes the snake forward. It can also coil up.

A snake has no arms, legs, shoulders, or hips. Its whole skeleton is a skull, backbone, ribs, and tail. A snake uses its many ribs as well as its backbone when it slithers along the ground.

Why can't you coil up like this python? The joints in your backbone aren't as flexible, and your arms and legs would get in the way.

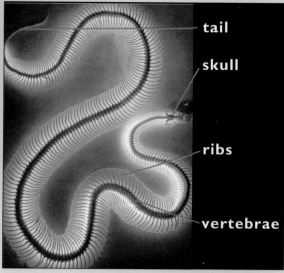

tail

skull

ribs

vertebrae

Large snakes like this python can have more than 400 vertebrae.

A leopard frog can leap more than 24 times the length of its body!

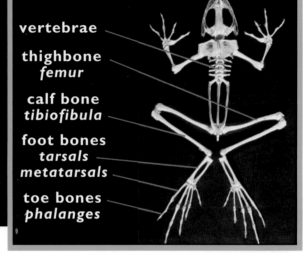

vertebrae

thighbone
femur

calf bone
tibiofibula

foot bones
tarsals
metatarsals

toe bones
phalanges

You have two bones in your lower leg, but a frog has just one strong bone.

A frog's skeleton is good for jumping. A frog can leap very far because its back legs are so much longer than its short backbone. Imagine how far you could jump if your legs were twice as long as your back!

When a frog gets ready to jump, it bends its hip, knee, and ankle joints very tightly. Then the long, powerful muscles attached to its hip, leg, and foot bones straighten the joints all at once and thrust the frog into the air. Its short front legs help it balance and act as shock absorbers when it lands.

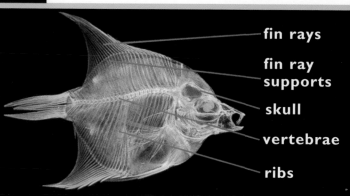

fin rays

fin ray supports

skull

vertebrae

ribs

This is an X-ray of an angelfish.

Angelfish have flat bodies that let them swim through tall, swaying water plants.

To swim, you use your arms and legs, but a fish has a different way of swimming. Like a snake, a fish depends on its flexible backbone. It moves forward by bending its body from side to side in a wavelike motion.

A fish also uses its tail and fins. A flick of its powerful tail can help a fish zoom ahead. Its fins help it turn left and right.

Some kinds of fish, such as sharks, have a skeleton made of **cartilage** instead of bone. Cartilage is a tough, rubbery material. You have it in your nose and ears and at the ends of your bones. It keeps the bones from scraping against each other in your joints as you move.

It's best not to "pick a bone" with a shark. Besides, it has no bones.

Fly and Glide

Birds have bones that are partly hollow and have thin walls. This makes their skeletons light enough for flying. The bones have supports on the inside to keep them from breaking.

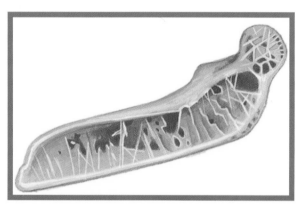

Bird bones have lightweight supports inside that look like webs.

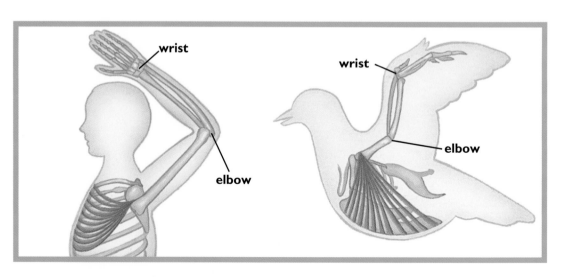

How are the bones in a bird's wing like the bones in your arm? How are they different?

This bald eagle is gliding with its huge wings stretched out. To fly higher, it will use its strong muscles to flap its wings.

Birds have large muscles attached to their breastbones at one end and to their wing bones at the other end. These bones and muscles work together when a bird flaps its wings.

Some of a bird's bones are **fused** together. They help form a rigid framework that protects the bird's heart and other organs from being crushed by its powerful flight muscles.

Animals with No Backbones

Many animals you know about are vertebrates. That is, they have backbones. However, most animal species have no backbones. Animals without backbones are called invertebrates. Ninety-seven percent of Earth's animal species are invertebrates.

Inside Out

Actually, invertebrates have no bones at all, but many of them do have skeletons. This group of animals includes insects, spiders, crabs, and lobsters.

Crabs and spiders, like insects, have jointed legs.

Your skeleton, like that of all vertebrates, is on the inside of your body. An insect's or a crab's skeleton is on the outside of its body. It's called an **exoskeleton**. Like a vertebrate's skeleton, an exoskeleton supports an animal's body and protects its organs. Your muscles are attached to the outside of your skeleton, but muscles are attached to the inside of an exoskeleton.

An insect's exoskeleton is made of a tough material called **chitin**. Unlike bone, an exoskeleton doesn't grow. From time to time as an insect grows, it **molts**. It grows a larger exoskeleton and pushes off the old one.

This cicada has just molted. It will leave its old exoskeleton on the twig.

19

No Skeletons at All?

Many invertebrates, such as earthworms and slugs, have no skeletons at all. Land animals that have no skeletons can't be very big because their bodies don't have much support. They also move slowly and not very far because they have no joints and their muscles are not attached to a skeleton.

Invertebrates that live in water can grow bigger than those that live on land because the water supports their bodies. If you have ever seen a jellyfish lying on the beach, you know that it looks like a shapeless blob. In the ocean, though, the water helps the jellyfish's body hold its shape.

To crawl, an earthworm tightens the muscles along the length of its body. How is this different from the way a human or an insect moves?

Moon jellyfish can be as big as 46 centimeters across. You can often see them on the sand after high tide or a storm.

Water even supports the body of a colossal squid, which can be more than fourteen meters long! The colossal squid is the largest living invertebrate—the largest animal with no bones.

Skeleton Clues

On land, animals must have bones to support a big body. We know about the biggest land animals that ever lived—dinosaurs like *Tyrannosaurus rex*—because people have found the **fossils** of their skeletons.

By studying an animal's skeleton, scientists can figure out what the animal probably looked like and how it moved. They can see how the bones fit together and where the muscles were attached.

Even millions of years after an animal lived, its skeleton can still tell an exciting tale.

From studying where muscles were once attached to bones, scientists know that *Tyrannosaurus rex* had extremely powerful jaws.

Glossary

cartilage: a stiff, rubbery tissue that is found at the ends of bones, in some other parts of the body, and in place of bones in some kinds of fish

chitin: the hard material that exoskeletons are made of

exoskeleton: a hard supporting structure on the outside of an animal's body. Animals such as insects, spiders, and crabs have exoskeletons.

fossil: the hardened remains or trace of an animal or plant that lived long ago

fused: joined in one piece, as if melted together

invertebrate: an animal without a backbone

joint: a place where two bones are connected

ligament: a strong tissue that connects bones in joints

marrow: the material inside bones in which new blood cells are formed

mineral: a natural substance that does not come from living things

molt: shed an outer covering, such as an exoskeleton. The old covering is replaced by a new one.

skeleton: the framework that supports and protects the body of an animal

species: a particular kind or type of living thing

vertebra: one of the separate bones that make up a backbone. Vertebrae means more than one vertebra.

vertebrate: an animal with a backbone

Index